Department of
Education

BUREAU OF NONPUBLIC SCHOOL REIMBURSABLE SERVICES

Title I
Middle School Activities

Key Curriculum Press
INNOVATORS IN MATHEMATICS EDUCATION

Writers: Dan Bennett, Christopher Casey, Greg Clarke, Larry Copes, Deidre Grevious, Lynn Hughes, Rhea Irvine, Ross Isenegger, Nick Jackiw, Tobias Jaw, Amy Lamb, Paul Kunkel, Ann Lawrence, Andres Marti, Daniel Scher, Nathalie Sinclair, Scott Steketee, Kelly Stewart, Kevin Thompson

Reviewers: Karen Anders, Susan Beal, Janet Beissinger, Dudley Brooks, Greg Clarke, Gord Cooke, Larry Copes, Judy Dussiaume, Paul Gautreau, Shawn Godin, Paul Goldenberg, Lynn Hughes, Scott Immel, Ross Isenegger, Sarah Kasten, Cathy Kelso, Amy Lamb, Dan Lufkin, Aaron Madrigal, Linda Modica, Margo Nanny, Henry Picciotto, Nicolle Rosenblatt, Joan Scher, Dick Stanley, Tom Steinke, Glenda Stewart, John Threlkeld, Philip Wagreich, Ken Waller, Bill Zahner, Danny Zhu

Field Testers: Laura Adler, Joëlle Auberson, Vera Balarin, Kim Beames, Judy Bieze, Caron Cesa, Heather Darby, Robin Glass, Cherish Hansen, Layne Hudes, Lynn Hughes, Scott Immel, Susan Friedman, Ann Lawrence, Vanessa Mamikunian, Michelle Mancini, Michelle Moore, Margo Nanny, Kelly O'Keefe, Tina Pierorazio, Leslie Profeta, Dechelle Rasheed, Cheryl Schafer, Joan Scher, Kimberly Scheier, JoAnne Searle, Char Soucy, Jessie Starr, Ruth Steinberg, Nancy Stevenson, Terry Suetterlein, Mona Sussman, William Vaughn, Ethan Weker, Angie Whaley

Module Editors: Elizabeth DeCarli, Karen Greenhaus

Activity Editors: Rhea Irvine, Daniel Scher, Josephine Noah, Scott Steketee, Cindy Clements, Joan Lewis, Silvia Llamas-Flores, Kendra Lockman, Lenore Parens, Glenda Stewart, Kelly Stewart

Production Editors: Angela Chen, Andrew Jones, Christine Osborne

Other Contributers: Judy Anderson, Elizabeth Ball, Tamar Chestnut, Brady Golden, Ashley Kuhre, Nina Mamikunian, Marilyn Perry, Emily Reed, Ann Rothenbuhler, Juliana Tringali, Jeff Williams

Copyeditor: Jill Pellarin

Printer: Lightning Source, Inc.

Executive Editor: Josephine Noah

Publisher: Steven Rasmussen

Key Curriculum Press
1150 65th Street
Emeryville, CA 94608
510-595-7000
editorial@keypress.com
www.keypress.com

ISBN: 978-1-60440-245-2
10 9 8 7 6 5 4 3 2 1 15 14 13 12 11

Contents

Sketchpad Resources

Sketchpad Learning Center

Help
Learning Center
Welcome Videos
Using Sketchpad
Teaching with Sketchpad
Reference Center
Objects
Tools
Menus
Online Resource Center
Sample Sketches & Tools
Check for Updates
Technical Support
Picture Gallery
License Information...
About Sketchpad...

The Learning Center provides a variety of resources to help you learn how to use Sketchpad, including overview and classroom videos, tutorials, Sketchpad Tips, sample activities, and links to online resources. You can access the Learning Center through Sketchpad's start-up screen or through the Help menu.

The Learning Center has three main sections:

Welcome Videos

These videos introduce Sketchpad from the point of view of students and teachers, and give an overview of the big ideas and new features of Sketchpad 5.

Using Sketchpad

This section includes 12 self-guided tutorials with embedded videos, 70 Sketchpad Tips, and links to local and online resources.

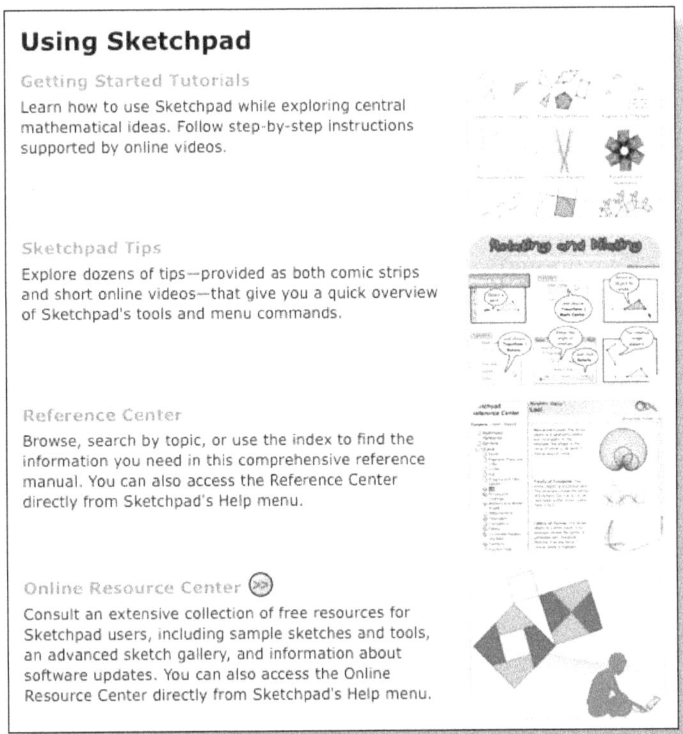

Teaching with Sketchpad

This section includes videos and articles describing how teachers make effective use of Sketchpad and how it affects their students' attitudes and mathematical understanding. There are over 40 sample activities, each with an overview, teaching notes, student worksheet, and sketches, that you can use with students to support your subject area, level, and curriculum.

Other Sketchpad Resources

Sketchpad contains resources for beginning and advanced users.

- **Reference Center:** This digital resource, which is accessed through the Help menu, is the complete reference manual for Sketchpad, with detailed information on every object, tool, and menu command. The Reference Center includes a number of How-To sections, an index, and full-text search capability.
- **Online Resource Center:** The Geometer's Sketchpad Resource Center (www.dynamicgeometry.com) contains many sample sketches and advanced toolkits, links to other Sketchpad sites, technical information (including updates and frequently asked questions), and detailed documentation for JavaSketchpad, which allows you to embed dynamic constructions in a web page.
- **Sketch Exchange:** The Sketchpad Sketch Exchange™ (sketchexchange.keypress.com) is a community site where teachers share sketches and other resources with Sketchpad users. Browse by keyword or topic for sketches that interest you, or ask questions and share ideas in the forum.
- **Sample Sketches & Tools:** You can access many sketches, including some with custom tools, through Sketchpad's Help menu. You can use some sample sketches as demonstrations, others to get tips and information about particular constructions, and others to access custom tools that you can use to perform special constructions. These sketches are also available under General Resources at the Sketchpad Resource Center (www.dynamicgeometry.com).
- **Online Courses:** Key Curriculum Press offers moderated online courses that last six weeks, allowing you to immerse yourself in learning how to use Sketchpad in your teaching. For more information, see Sketchpad's Learning Center, or go to www.keypress.com/onlinecourses.
- **Other Professional Development:** Key Curriculum Press offers free webinars on a regular basis. You can also arrange for one-day or three-day face-to-face workshops for your district or school. For more information, go to www.keypress.com/pd.

Addressing Grade-Level Learning Objectives

The table below shows how the activities in this collection align to the Title I Mathematics Learning Objectives for Grades 6–8. The Activity Notes for each activity guide you in using dynamic models, manipulatives, and constructions to model and explore mathematical concepts with your students, in leading class discussions, and in helping students explain and justify their solutions using appropriate mathematical terminology.

Activity Title	Learning Objective
Chapter 1: Number Comparisons and Properties in Grade 6	
Shape and Size: Exploring Similar Triangles	Interpret and write ratios in different contexts to show the relative sizes of two quantities, using appropriate notation (*a*/*b*, *a* to *b*, *a*:*b*) Express equivalent ratios as a proportion, and use proportions to solve problems (find the length of a side of a polygon similar to a known polygon); use cross-multiplication as a method for solving such problems
Stretchy Percent Ruler: Percent Concepts	Read, write, and identify percents of a whole (0%–100%) Find the percent of a quantity as a rate per 100 (30% of a quantity means 30/100 times the quantity); solve percent problems involving finding the whole, given a part and the percent
Jump Along: Positive and Negative Integers on the Number Line	Order positive and negative numbers, fractions, decimals, and mixed numbers, and place/locate them on a number line
Right or Left: Adding and Subtracting Integers	Define and demonstrate an understanding of the absolute value of rational numbers as distance from zero on the number line ($\|-4\| = 4$, $\|7\| = 7$, and $\|-3.9\| = 3.9$) Order positive and negative numbers, fractions, decimals, and mixed numbers, and place/locate them on a number line
Jeff's Garden: Area Model of Fraction Multiplication	Add and subtract fractions with unlike denominators; multiply and divide fractions by fractions; simplify/reduce fractions to lowest terms
Feed the Mouse: Fraction and Decimal Relationships	Compare, order, estimate, find equivalence, and translate among percents and rational numbers, including positive and negative integers, fractions, mixed numbers, and decimals ($0.32 \approx 1/3$, and $0.32 = 32\%$)
Algebars: Exploring Order of Operations	Read, write, and evaluate a numeric expression, and an algebraic expression for a given situation, using up to three variables ($2a + b = c$) Apply algebraic order of operations and the commutative, associative, and distributive properties to evaluate expressions

Activity Title	Learning Objective
Chapter 2: Shapes, Graphs, and Data in Grade 6	
Slanted Bases: Calculating Triangle Area One Parallel Pair: Trapezoid Area Smoothing the Sides: Regular Polygon and Circle Area	Evaluate formulas for given input values (circumference, area, volume, distance, temperature, interest, etc.) Develop formulas and use various strategies to find the area of a right triangle, other triangles, regular and irregular polygons
Stack It Up: Volume of Rectangular Prisms Perfect Packages: Surface Area and Volume	Find the volume of a right rectangular prism, triangular prism, and cylinder
Fly on the Ceiling: Coordinate Systems	Graph/plot points and identify coordinates in all four quadrants on the Cartesian coordinate plane; draw polygons in the coordinate plane given coordinates for the vertices
Wavy Parallelogram: Circle Area	Identify, measure, and describe circles and the relationships among a circle's radius, diameter, circumference, and area; recognize pi (π, 3.14,) as a ratio of circumference to diameter; use formulas to solve problems
Making Means: Data Distribution and Averages Mean Meets the Median: Measures of Central Tendency Quartile Craze: Box-and-Whisker Plots	Determine the mean, median, mode, and range of a given set of data Read, interpret, and display numerical data in various graphs; plot on a number line (including dot and box plots), histogram, bar/line/circle graph, picture graph
Chapter 3: Numbers, Expressions, and Equations in Grade 7	
Open the Safe: Multiples and Factors	Find the common factors, the greatest common factor, and least common multiple of two or more numbers
The Envelope: Adding and Subtracting Integers	Add and subtract integers with and without the use of a number line ($+7 - (-5) = 12$)
Right or Left: Adding and Subtracting Integers	Understand and show that a number and its opposite have a sum of 0 (are additive inverses); interpret sums of rational numbers by describing real-world contexts Add and subtract integers with and without the use of a number line ($+7 - (-5) = 12$)
Paper Cups: Connecting Slope and Unit Price	Express ratios in several ways (3 cups to 5 people; 3:5; 3/5) and recognize equivalent ratios Calculate and compare unit rates/prices using proportions Use proportional relationships to solve multi-step ratio and percent problems (simple interest, tax, markups, markdowns, gratuities and commissions, fees, percent increase and decrease, percent error)
Signed Tiles: Evaluating Expressions	Use correct order of operations to evaluate algebraic expressions that include variables and/or integral exponents greater than zero
Balancing: Solving Linear Equations Balancing with Balloons: Solving Equations with Negatives Undoing: Solving Linear Equations	Solve multi-step equations, such as $2n + 3 + 5n = 10$ by combining like terms

Activity Title	Learning Objective
Chapter 4: Geometry, Data, and Probability in Grade 7	
Prism Dissection: Surface Area Pyramid Dissection: Surface Area	Use formulas to calculate the surface area and volume of three-dimensional solid figures in order to solve word problems
Dilation Designs: Proportions in Similar Polygons Shape and Size: Exploring Similar Triangles	Differentiate between congruent and similar figures and angles, and apply these relationships to solve problems
Double Cross: Angles Formed by a Transversal	Use facts about supplementary, complementary, vertical, and adjacent angles to find a missing angle in a two-dimensional figure
Squaring the Sides: The Pythagorean Theorem	Understand and apply the Pythagorean Theorem ($a^2 + b^2 = c^2$) to determine the unknown length of a side of a right triangle
Apartment Floor Plan: Scale Drawing	Solve problems involving similar figures, scale drawings and maps; include computing actual lengths and area from a scale drawing; and modify a scale drawing proportionally
Point Graphs: Representing Data	Read, display and interpret data represented graphically (number line, dot plots, histogram, line graph, double line/bar graphs, circle graphs, or box plots)
Dartboards: Geometric Probability	Use random sampling to draw inferences about a population; investigate chance processes; develop, use, and evaluate probability models
Chapter 5: Exponents and Polynomials in Grade 8	
Powering Up: Multiplication and Exponents Powering Down: Division and Exponents Power Strips: Laws of Exponents	Understand, apply, and explain the laws of exponents (division: $a^5/a^2 = a^3$; multiplication: $(a^2)^3 = a^6$, or $(a^2)(a^5) = a^7$) Evaluate algebraic expressions with integral exponents ($y^3 = y \cdot y \cdot y$)
Broccoli and Brussels Sprouts: The Distributive Property Tiling in a Frame: Multiplying Polynomials Tiling Rectangles: Factoring Polynomials Dynamic Tiles: Evaluating Polynomial Expressions	Factor algebraic expressions using the greatest common factor (GCF) ($x^2 + 3x + 2 = (x + 2)(x + 1)$) Multiply a binomial by a monomial, such as $(2x + 3)(4x)$, or a binomial by a binomial, such as $(4x + 1)(2x^2 + 3)$
Chapter 6: Geometry and Graphs in Grade 8	
Mellow Yellow: Interpreting Graphs	Describe a situation involving relationships that matches a given graph
Shady Solutions: Graphing Inequalities on a Number Line	Solve multi-step inequalities and graph the solution set on a number line
Amazing Angles: Finding Transversal Angle Pairs	Calculate the missing angle in a supplementary or complementary pair Identify the angle pair relationships and calculate the missing angle measurements when given two parallel lines cut by a transversal

Activity Title	Learning Objective
Chapter 6: Geometry and Graphs in Grade 8, ***continued***	
Menagerie: Comparing Transformations Transformers: Exploring Coordinate Transformations	Identify, describe, and draw geometric transformations in the plane, using proper function notation (rotations, reflections, translations, and dilations)
Rainbow of Lines: Investigating Slope and Intercepts Dance Pledges: Plotting Linear Equations Hikers: Solving Through Multiple Representations	Write an equation of a line in slope/intercept form ($y = mx + b$) Determine the slope of a line given the coordinates of two points on the line, and explain the meaning of slope as a constant rate of change Explain the meaning of the y-intercept of a line, and determine the y-intercept of a line from a graph and from an equation of the line Analyze, solve, and graph linear equations and pairs of simultaneous linear equations graphically, algebraically, and from a table of values
Cell Phone Plans: Comparing Graphs	Determine the slope of a line given the coordinates of two points on the line, and explain the meaning of slope as a constant rate of change Read and interpret data represented graphically (including picture/bar/line graph, histogram, double bar/line graph, circle graph, and/or scatter plot)

1

Number Comparisons and Properties in Grade 6

Shape and Size: Exploring Similar Triangles

ACTIVITY NOTES

INTRODUCE

Project the sketch for viewing by the class. Expect to spend about 5 minutes.

1. Open **Shape and Size.gsp.** Go to page "Similar Triangles." Enlarge the document window so it fills most of the screen.

2. Explain, ***Today you're going to use Sketchpad to investigate the properties of similar triangles. What does it mean when two figures are similar?*** Take students' suggestions. If they only say that the two figures have the same shape, ask students whether the figures must be the same size. ***Do the two figures have to be the same size?*** Write an agreed-upon definition on the board: *Two figures are similar if they are identical in shape, but not necessarily the same size.* You might want to clarify that similarity requires more than just the same general shape: Not all triangles are similar. For your understanding, this activity does not consider reflections that are similar.

3. Direct students' attention to the sketch. Press *Show Vertices.* ***What are the names of these two triangles?*** [$\triangle ABC$ and $\triangle DEF$] If needed, review how to name triangles by their vertices. Drag vertices D and E on the blue triangle. ***What do you notice?*** Students may note that the position, orientation, and size of the blue triangle changes, but its shape and the sizes of its angles do not change. (The term *orientation* is used informally in this activity to refer to rotation, not the order of vertices.) Don't press students if they miss some observations; they will investigate the properties of similar triangles further during the activity.

4. ***In this activity you'll explore what makes these two triangles similar. You'll learn what the properties of similar triangles are and use them to find missing side lengths.***

DEVELOP

Expect students at computers to spend about 30 minutes.

5. Assign students to computers and tell them where to locate **Shape and Size.gsp.** Distribute the worksheet. Tell students to work through step 22 and do the Explore More if they have time. Encourage students to ask their neighbors for help if they are having difficulty with Sketchpad.

6. Let pairs work at their own pace. As you circulate, here are some things to notice.

   - In worksheet step 4, before students press *Blue to Yellow,* have them make the blue triangle dramatically smaller and in a different orientation than the yellow triangle. ***What happens during the***

animation? Help students identify the three stages of the animation: change in position (translation), change in orientation (rotation), and change in size (dilation). Students can also press *Yellow to Blue* to see the yellow triangle translate, rotate, and dilate.

- In worksheet step 6, as students are dragging the vertices of each triangle, ask them to try to predict which vertex of the blue triangle corresponds to each vertex of the yellow one. ***Can you tell which angle in the blue triangle corresponds to angle A in the yellow triangle? How about angle B? Angle C?*** Be sure students understand the term *corresponds* and define it as "matches," if necessary. Students can check their answers in the next step when they match up the triangles.
- In worksheet step 7, if students have trouble overlaying the triangles exactly, have them first press *Align Triangles* and then overlay the blue triangle on the yellow triangle.
- In worksheet step 9, review how to name angles by three points: a point on one side, the vertex point, and a point on the other side. Remind students that the vertex point must always be the second point named.
- In worksheet step 12, have students drag all the vertices. Encourage them to make triangles of different sizes. ***What happens to the angle measurements if you make the blue triangle really big? What happens to the angle measurements if you make the blue triangle really small? Try making the yellow triangle's angles all equal. What do you notice about the angle measures in the blue triangle?*** If necessary, remind students that the symbol $m\angle$ means "measure of the angle" and that angle measurements units are degrees.
- In worksheet step 14, students can choose the **Segment** tool and then choose **Edit | Select All Segments** to select all the segments in both triangles before measuring the lengths. After students measure the side lengths, ask them to drag the vertices and watch what happens to the measurements. ***If you drag the vertices of the blue triangle to change its side lengths, what happens to the side lengths in the yellow triangle?***
- In worksheet step 15, review the definition for *ratio* as needed: *A ratio is a comparison of two quantities by division.* Then check that students understand what the corresponding sides are in the two triangles.

Which side in the yellow triangle corresponds to side AB in the blue triangle? What other sides in the two triangles correspond?

- An alternate method for finding the ratio of side lengths is to select two sides (not their measure lengths) and choose **Measure | Ratio.** This would accomplish the same result as worksheet steps 14 and 15.
- In worksheet step 16, after students calculate all three ratios, have them drag the vertices of the triangles and observe what happens to the ratios. ***What do you notice about the ratios?*** Students should observe that the ratios change, but all three remain equal to each other.
- In worksheet step 18, remind students that a proportion is made up of two equal ratios. Tell students it's not fair to interchange the left and right sides of a proportion and count it as different.
- In worksheet step 21, students may find that their answers differ from other students' answers due to Sketchpad's rounding.
- If students have time for the Explore More, they will use similar triangles to find the length of a bridge. If students are stumped, help them identify the two similar triangles they can use. ***What triangle do you see in the scale drawing? How can you find the length of the two sides shown? What similar triangle can you use in the real world? What measurements do you know? What measurement are you trying to find out?***

SUMMARIZE

Project the sketch. Expect to spend about 10 minutes.

7. Gather the class. Students should have their worksheets with them. Begin the discussion by opening **Shape and Size.gsp** and going to page "Similar Triangles." ***What did you learn about the properties of similar triangles?*** Let volunteers come up to the computer and demonstrate. Students should respond that in similar triangles the corresponding angle measures and the ratios of corresponding side lengths are equal.
8. Then have students demonstrate how they solved the problems in worksheet steps 20 and 22. ***What proportion did you use to solve the problem? How did you know how to set up the proportion?*** Students should describe how they set up ratios of corresponding sides.

9. ***Suppose the ratio of corresponding sides of similar triangles is one to one. What do you know about the triangles?*** Students should recognize that the triangles are congruent. Demonstrate on the sketch for the class.

10. If time permits, discuss the Explore More. Ask students to describe the two similar triangles and what proportion they used to solve the problem. Triangle *ABC* in the scale drawing is similar to the triangle *ABC* in the real world. Students need to measure the lengths of the sides in the sketch to set up a proportion equivalent to $\frac{4.42}{x} = \frac{2.49}{500}$. Check that students understand that the answer is in meters.

11. ***Are all right triangles similar? Explain.*** You may wish to have students respond individually in writing to this prompt. Right triangles might have only one angle that is the same measure—the right angle. The two other angles measures may not be equal, so not all right triangles are similar.

EXTEND

What questions occurred to you about similar figures? Encourage curiosity. Here are some sample student queries.

What if you flipped one of the triangles over?

Is the ratio of two sides in one triangle the same as the ratio of the corresponding two sides in a similar triangle?

We just looked only at triangles. Do these properties hold true for other similar figures?

What other types of problems can you solve using similar triangles?

ANSWERS

3. As you drag points *D* and *E*, the angles of the blue triangle stay the same, but its size, position, and orientation change.

4. Both buttons show an animation that suggests the two triangles are the same shape.

7. The triangles match up when point *D* is on top of *A*, and *E* is on top of *B*.

9. When you make $\angle CAB$ in the yellow triangle very small, $\angle FDE$ in the blue triangle also becomes small.

10. When you drag point C far away from points A and B, point F moves far away from points D and E.

12. Angle DEF in the blue triangle corresponds to $\angle ABC$ in the yellow triangle, and $\angle EFD$ corresponds to $\angle BCA$. Angle CAB in the yellow triangle corresponds to $\angle FDE$ in the blue triangle.

13. The corresponding angles of similar triangles are equal in measure (or congruent).

17. The ratios of the three pairs of corresponding side lengths of similar triangles are equal.

18. Students should write these three proportions or their equivalents:

$$\frac{m\overline{AB}}{m\overline{DE}} = \frac{m\overline{CA}}{m\overline{FD}} \qquad \frac{m\overline{AB}}{m\overline{DE}} = \frac{m\overline{BC}}{m\overline{EF}} \qquad \frac{m\overline{BC}}{m\overline{EF}} = \frac{m\overline{CA}}{m\overline{FD}}$$

19. Corresponding angles of similar triangles are equal in measure (or congruent), and the ratios of corresponding side lengths are equal.

20. $m\overline{EF} = 6.7$ cm. This result is rounded off to the nearest tenth, as are all the distances in this activity.

21. Because of rounding, some students may end up with slightly different answers, such as $m\overline{EF} = 6.6$ cm.

22. $m\overline{AB} = 2.6$ cm. The proportion is $\frac{m\overline{AB}}{m\overline{DE}} = \frac{m\overline{CA}}{m\overline{FD}}$.

23. There are two similar triangles: One triangle appears in the scale drawing, with sides that students should measure in centimeters. The other straddles the banks of the river itself, with one side of 500 m and another side that is the unknown length of the bridge. One approach is to measure the distances in the scale drawing (using centimeters) and to set up a proportion involving the corresponding sides of the two triangles. The bridge must be 888 m in length.

Shape and Size

Name:

Two geometric figures are *similar* if they have the same shape, but not necessarily the same size. In this activity you'll investigate the properties of similar triangles and use proportions to find the missing sides of a pair of similar triangles.

EXPLORE

1. Open **Shape and Size.gsp** and go to page "Similar Triangles."

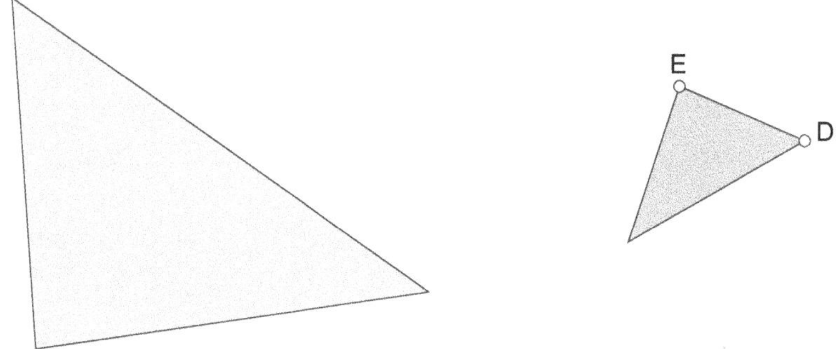

2. Drag points *D* and *E* to change the blue triangle.

3. As you drag the points, what changes and what stays the same?

4. Press *Blue to Yellow.* Then press *Yellow to Blue.* What does this seem to show about the two triangles?

5. To change the shape of the yellow triangle, first press *Show Vertices.*

6. Drag the vertices of each triangle. Observe what changes and what stays the same.

7. Drag point *D* and *E* until $\triangle DEF$ lies exactly on top of $\triangle ABC$. Where do you have to place points *D* and *E* to make the triangles match up?

8. Drag the triangles apart again.

9. Make $\angle CAB$ in the yellow triangle very small. What happens in the blue triangle?

10. Drag point *C* far away from points *A* and *B*. What happens in the blue triangle?

11. Press *Show Angle Measurements*.

12. Drag each vertex to see what happens to the angle measurements. Which angle in the blue triangle corresponds to $\angle ABC$ in the yellow triangle? Which angle corresponds to $\angle BCA$? Which angle in the yellow triangle corresponds to $\angle FDE$?

13. What can you conclude about the corresponding angles of similar triangles?

14. Select each side and choose **Measure | Length.**

15. Calculate the ratio of one pair of corresponding sides. To do so, choose **Number | Calculate** to open the Sketchpad Calculator. Click on a measurement to enter it into a calculation.

16. Calculate the ratios of the other two pairs of corresponding side lengths.

17. What can you conclude about the ratios of the corresponding side lengths of these triangles?

18. Complete the following proportion using the ratios you calculated in steps 15 and 16. Then write two other proportions using these ratios.

$$\frac{m\overline{AB}}{\phantom{m\overline{DF}}} = \frac{\phantom{m\overline{AB}}}{m\overline{DF}}$$

19. Using your answers to steps 13 and 17, write a summary of the properties of similar triangles.

20. If $m\overline{AB} = 3.0$ cm, $m\overline{BC} = 4.0$ cm, and $m\overline{DE} = 5.0$ cm, use the triangles to find $m\overline{EF}$. Set up a proportion to check your answer.

21. Compare you answer to step 20 with other students' answers. Then compare your sketches. Explain any similarities or differences.

22. If $m\overline{CA} = 4.5$ cm, $m\overline{DE} = 2.0$ cm, and $m\overline{FD} = 3.5$ cm, use the triangles to find $m\overline{AB}$. Set up a proportion to check your answer.

EXPLORE MORE

23. Go to page "Explore More." This sketch shows a scale drawing of a river and the location of a bridge that must be built between point *A* and point *B*. There is no easy way to directly measure the distance between the two points on opposite sides of the river. Use similar triangles and proportions to find the required length of the bridge.

Stretchy Percent Ruler: Percent Concepts

ACTIVITY NOTES

INTRODUCE

Project the sketch for viewing by the class. Expect to spend about 15 minutes.

1. Open **Stretchy Percent Ruler Present.gsp.** Enlarge the document window so it fills most of the screen. Begin by introducing the ruler.

   Go to page "Ruler 1." Explain, ***You're going to be using an unusual ruler. Let's get to know it. What do you notice? How is this ruler like rulers you've used to measure things?*** Elicit the ideas that the tool is linear ("straight") and is marked in intervals of equal size. Students are likely to observe that, unlike on a ruler, there are no number labels for the intervals.

   Using the **Arrow** tool, drag the point at the top of the ruler upward to enlarge the ruler and then downward to shrink it. ***What's different about this ruler?*** Here are several sample responses.

   A normal ruler stays the same length. This ruler gets larger or smaller depending on how you drag it.

   A normal ruler has marks that divide it into equal parts. This one does too, but the distance between the marks changes as the ruler grows or shrinks.

   This ruler is divided into four equal intervals. No matter what you do, the intervals always divide the ruler evenly into four parts.

   Go to page "Ruler 2." Drag the ruler's endpoints, one at a time. ***What can you say about the ruler now?*** Elicit these ideas: The intervals are labeled; the labels are in percents rather than a unit of length, such as inches or centimeters; and again, enlarging or shrinking the ruler maintains the relationship of the intervals to the whole (50% is always 50% of the ruler's length).

Students will enjoy learning that giraffes have the same number of neck vertebrae (seven) that humans have.

2. Go to page "Giraffe." ***Here's a picture of a giraffe that's drawn to real-life proportions. Suppose we want to know how the length of a giraffe's head and neck compares to its total height. How could we use the percent ruler and this picture of a giraffe to find out?*** Draw on students' informal knowledge to clarify the meaning of *real-life proportions.* A student might explain, *It means that in the drawing, the relationship of the parts of the giraffe to each other and to the whole giraffe are the same as the way those things are in real life. If the head is a tenth of the height in the drawing, then it's a tenth of the height in real life.*

Point to the position marked by the horizontal line in the illustration at right and let students know that this is where the bottom of the last vertebra of a giraffe's neck is located.

Have students visually estimate the percent and share their estimates.

Invite a volunteer to the computer to experiment with the tool. Students may be inclined to position 0% at the bottom of the giraffe or at the top, and to position the ruler alongside or over the drawing. Each of these approaches may be appropriate. Expect discussion of whether the horns are part of the head and where the feet "are standing." Decide as a class that because students are looking for an estimate, decisions about the top and bottom of the giraffe (and differences of opinion) are not critical.

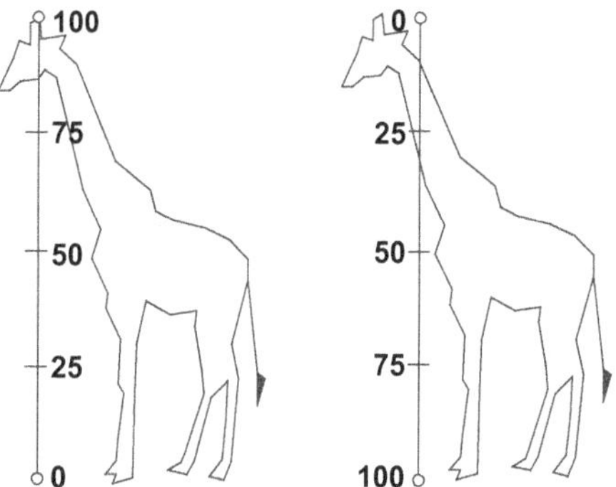

About what percent of the giraffe's height is its head and neck? The class is likely to agree on an estimate such as "a little less than 50%." This may be similar to the estimate students arrived at visually.

3. ***Let's find the percent of the giraffe's leg that is below the knee. What's your estimate?*** Take responses. Invite a volunteer to the computer to use the stretchy ruler and check students' estimates. Elicit agreement that about 50% of the giraffe's leg is below the knee.

Students will need to be alert to the different "wholes" in the guidelines for facial proportions they will be using.

Check students' understanding of the role of the "whole" in relation to a percent. ***The part of the leg below the knee is about 50% and the neck and head are about 50%. Are these about the same length?*** This student response shows understanding: *Because the things that are 100% are different sizes, 50% has to be different amounts. The height of the giraffe is greater than the length of the whole leg, so we know that 50% of those is going to be different.*

4. ***We said that the length of a giraffe's head and neck is a little less than 50% of a giraffe's height. Is that true for people?*** Have students pair up and take turns standing so partners can estimate what percent of each partner's height is the length of his or her head and neck. (Alternatively, the class might suggest having several students stand at the front of the room.)

   What do you think? Here are two sample responses.

   We think that for people, it's about 20%. It's less than a quarter of the total height, and that would be 25%.

   We think it's about halfway between 0% and 25%, so it's about 12% or 13%.

5. Go to page "Man." ***This image of an adult man is drawn to real-life proportions.*** Have a volunteer come to the computer to use the ruler and find the approximate percentage that is the head and neck.

   Now have the volunteer press *Show 10% Ruler.* Have the class describe the ruler, which is marked in intervals of 10%. ***Would it be helpful to use this ruler? Why?*** Note whether any students think this ruler may allow them to make a closer (more precise) estimate because the intervals are smaller. Have the volunteer try the ruler. Here is a sample response. *Now I'd say that the head and neck are about 15% of the height of a human—not more than 15%.*

6. Go to page "Draw a Face." Distribute the worksheet. Explain that students will draw a face using guidelines that artists use for facial proportions. ***What do we mean by* facial proportions*?*** [The relationship of parts of the face to each other and to the whole head] Explain that because individuals are different, artists adjust the proportions when they draw, paint, or sculpt real humans; today students are going to use the given proportions, without adjusting them.

7. Read together the guidelines in worksheet step 2. Students will notice that some of the guidelines are given in fractions. ***How do you think you'll use one or the other ruler with these fractions?***

   The eyes are half the length, so I'll fit the ruler to the length of the head and then see where 50% on the ruler is, because 50% is half.

   One-third is a little more than 33% (100 divided by 3), so I would use the 10% ruler and go a little more than 30%.

8. Model just enough to prepare students to work on their own.

   - Using the **Compass** tool, draw a circle for the head, and point out that this circle is *the head* or *the face* in the guidelines. Explain that although heads are not truly circular in shape, for this activity a circle will be used because it is easy to construct.

   - Drag the point on the circle and then drag its center to demonstrate that dragging either one will change the size of the circle. ***When you are sure of the size circle you want, select both points and choose*** **Display | Hide Points.** Model this. ***Why is this important?*** Elicit the idea that if students were to change the size of the circle while they were drawing the rest of the face, the relationship of the parts of the face to the whole face would change—the facial proportions would change.

Don't suggest a circle size. Your hope is that students draw circles of recognizably different sizes, as this will be useful in discussion later.

   - Make a nose by drawing several connected segments. Using the **Arrow** tool, create a selection rectangle to select the entire nose. Then drag the nose to another location on the face. Students will need to create a selection rectangle if they want to drag any facial features made by drawing more than one object (segment or circle).

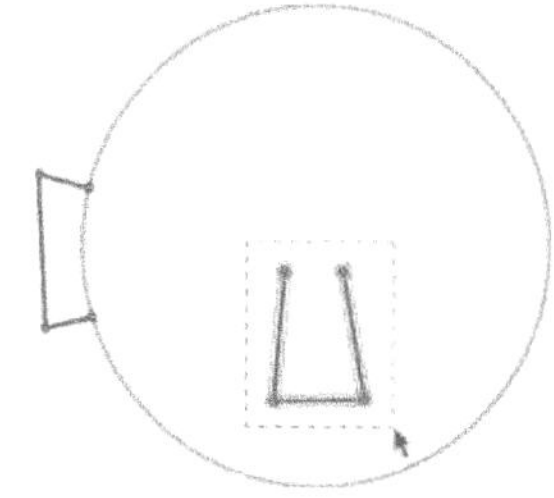

See page "Draw a Face" for an example.

   - Point out that there is no guideline for how far the hairline is from the top of the head. Clarify that the hairline is the place where hair starts growing at the top of the face. Students will use their own judgment to locate the hairline on the faces they draw, use the guideline to determine the location of the eyebrows, and then draw hair (if they wish) in any way they like.

   - Using the **Segment** tool, draw an ear on the circle you've drawn. Point out that students should make sure the circle is highlighted when they move the pointer over the circle to begin the ear; this ensures that the segment and the circle are connected. If students drag the circle (which they should not need to do), the shape and size of the ear will change; students will need to drag parts of the ear to make adjustments.

- As they work with guidelines 2d and 2f, students will need to drag a ruler while maintaining its length. Demonstrate selecting the segment, rather than one of its endpoints, to do this.
- Model other tools and commands in worksheet step 2 if they are new to your students.

9. If you want students to save their work, demonstrate choosing **File | Save As,** and let them know how to name and where to save their files. If students will print their faces, model choosing **File | Print Preview**; in the dialog box that appears, set the image to fit on one page before clicking **Print.**

DEVELOP

Expect students at computers to spend about 45 minutes.

10. Assign students to computers and tell them where to locate **Stretchy Percent Ruler.gsp.** Students should have their worksheets. Tell students to work through step 2 and do the Explore More if they have time. Read the Explore More together and be sure students understand what they will be doing. Encourage students not to spend so much time drawing the parts of the first face that they don't have time to draw a funny face. Encourage students to ask their neighbors for help if they have questions about using Sketchpad.

11. Let pairs work at their own pace. As you circulate, here are some things to notice.

    - At the start, remind students to hide the point at the center of the circle and the point on the circumference when they have decided on the size of the face. Also, if a student pair draws a small circle for the head, don't suggest they draw a larger circle unless you think they will have trouble fitting the facial features. A range of circle sizes will be needed for the class discussion later.
    - For most students, the easiest way to draw the parts of the face will be to follow the guidelines in the order they are given.
    - Ask students to explain how they know the proportion is correct for the facial feature they are drawing.
    - For guidelines 2a and 2b, a good strategy (and one students are likely to use) is to align the endpoints of a ruler with the top and bottom of the circle through the center of the face.

- For guideline 2c, the "whole" is no longer the whole face; it's now the distance from the chin to the bottom of the nose. Observe how students work with this change.
- For guideline 2d, remind students they do not need to draw a hairline. Also remind them, as necessary, that the hairline is where the hair grows across the top of the brow, not at the top of the head.
- For guideline 2d, students are likely to need to think about how to use the ruler. Allow them to struggle with this interesting problem. If you need to, offer prompts that keep students doing their own thinking. ***You know how to show the length of the face on your ruler.*** [Students are likely to align the endpoints with the top and bottom of the circle through the center of the face. Make sure students then drag the ruler, maintaining its length, outside of the face to continue.] ***And what does guideline 2d say about how far the eyebrows are from the hairline?*** [One-third the length of the face] ***How can you use your ruler to tell how far that is?*** [Find approximately 33% on the ruler.] ***How can you use that to figure out where the eyebrows should go?*** [Drag the ruler so 0% is at the hairline and draw the eyebrows at approximately 33% on the ruler.] Students will use the same process to draw the length of the ear (guideline 2f).
- Students are likely to be interested in the faces other pairs are creating. Some stimulating observations may come of this. They may, for example, notice another pair's drawing in progress and observe that the size of the head differs from the size in their own drawing. Note whether students observe that this doesn't mean the facial proportions change. You might hear a comment such as, *Their head is a lot smaller than ours, but the eyes are still in the middle, 50% of the way from the top. And the mouth looks right.*

This big idea will be discussed later by the class.

- Note any students for whom using benchmark percents and fractions interchangeably is a secure skill.
- As students are finishing work on page "Draw a Face," make note of faces you want to share with the class in the discussion that follows. Be sure to identify two drawings that have circles of different sizes for the head and the correct facial proportions.
- If students have time for the Explore More, make sure they are recording the proportions they are using. You might prompt,

In your funny face, what percent of the face is the distance from the top of the head to the eyes?

12. If you have a flash drive available, collect sketches to display from the shared computer. Otherwise, remind students how to print on one page and have them print now. If students will print their work, they might wish to hide the points first. Provide these directions. (Note that this will cause the endpoints of the rulers to be hidden as well.)

    - Select the **Point** tool.
    - Choose **Edit | Select All Points.**
    - Choose **Display | Hide Points.**

    If students will save their work, let them know how to name and where to save their files now.

SUMMARIZE

Project student sketches, if possible. Expect to spend about 15 minutes.

13. Gather the class. Students should have their worksheets with them. Show two faces students have drawn that have very different size circles for the faces and follow the facial proportions guidelines. Have students confirm that both faces follow the guidelines. If students need more than a visual inspection, take the time to use the rulers to check both drawings.

    (To project two faces, you might copy one face and paste it onto the sketch page containing the other face. Use a selection rectangle to select the face, choose **Edit | Copy,** click in the other sketch, and choose **Edit | Paste;** with the entire face selected, drag it where you want it.)

14. Facilitate a discussion in which students compare the distances for facial features on one face and on the other. Look for evidence that students are able to distinguish an absolute comparison (such as the comparison of the actual distances from the chin to the mouth) and a relative comparison (the relationship of those distances to the lengths of the respective faces). The goal is for students to be convinced that "the distances can be different but the relationships are the same." Here are some suggested questions to pose.

    Let's compare these faces. What's different about the distances between parts of the face on this head and the distances on this head? [The distances are greater on one face, less on the other.]

What's the same about the distances on both faces? [The relationships of the distances to each other and to the whole head]

Please give me an example. A sample response is, *The distance from the chin to the bottom of the nose in both drawings is 25% of the face's length. That's the same on both faces. But the actual distance is longer from the chin to the nose on the face on the right.*

How can the distance in both drawings be 25% of the whole length, but the distances turn out not to be the same? Like the response above, an explanation that shows good understanding of the concept of the unit as well as the student's ability to think in terms of a relative comparison is this one: *It depends on the size of the whole face. Twenty-five percent of our face is not the same distance as 25% of their face because the size of the faces is different.*

15. ***How would you explain to someone else how a percent ruler is different from a regular ruler?***

    A percent ruler shrinks and grows because, for example, 50% isn't an exact amount like 3 inches is an exact amount. So you can't mark 50% at exactly one place like on a regular ruler.

    The percent ruler needs to stretch or shrink to be the size of the whole thing you want to find a percent of. You have to know what 100% is, and then you can find other percents.

    Fifty percent is 50% of something. And that thing can be big or small, so 50% can be big or small. It just depends on how much the whole thing is.

16. If students have drawn "funny faces" in Explore More, display as many as time permits. For each sketch, invite the class to estimate visually one or two of the proportions used and observe as you check using a ruler.

EXTEND

1. ***What questions about percents and proportions occur to you?*** Encourage curiosity. Here are some sample student queries.

    How wide is a person's mouth, usually, in relation to the width of the face or in relation to the width of an eye?

    Is the distance of the mouth across the face 100% of the distance from one pupil to the other? Those seem about the same.

If you smile really big, can your mouth (from side to side) be 100% the width of your eyes?

Are body proportions the same for people of all ages? Are facial proportions the same or different?

What proportion is a human baby's head and neck compared to its height? What about a baby giraffe's?

I'd like to look at different artists' work and see how much they adjust the guidelines.

Is there a reason to use percents instead of fractions?

2. Human faces are roughly symmetric. Have students use the artists' guidelines to draw half a face, and then use reflection to complete the face.
   - Draw a circle for the face and place a point on the circle at the top of the head. With that point and the circle's center selected, choose **Construct | Line.** Choose **Display | Line Style | Dashed.** This will be the line of reflection.
   - Following the guidelines, draw half the face.
   - Using the **Arrow** tool, select the line of reflection and choose **Transform | Mark Mirror.** The line will flash briefly to indicate it has been marked as the mirror.
   - Using a selection rectangle, select all the facial features (this will select the circle for the head too) and choose **Transform | Reflect.**

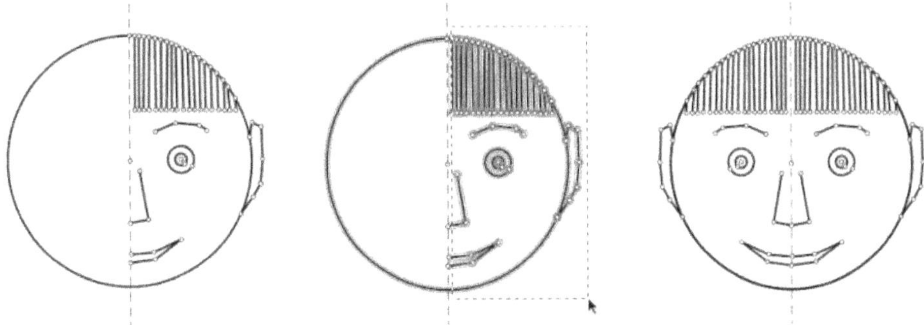

3. In this activity students were not given any proportions for the width of facial features. Have them research the proportions that artists use.

4. Some students may be curious about how the percent ruler was constructed. On page "Make Your Own" of the student and teacher sketches, directions are given for making a basic stretchy ruler.

ANSWERS

2. Check students' drawings. Distances will vary depending on the size of the circle used for the head, but proportions should match the guidelines.
3. Check students' drawings. Proportions used in the drawings should not match the guidelines; they should match what students have recorded.

Stretchy Percent Ruler

Name:

Draw a face using the guidelines for proportions that artists use.

EXPLORE

1. Open **Stretchy Percent Ruler.gsp.** Go to page "Draw a Face."

2. Follow these guidelines as you draw a face. The length of the face is the distance from the top to the bottom of the head.
   a. The distance from the top of the head to the eyes is one-half the length of the face.
   b. The distance from the chin to the bottom of the nose is 25% of the length of the face.
   c. The distance from the chin to the mouth is 50% of the distance from the chin to the bottom of the nose.
   d. The distance from the hairline to the eyebrows is one-third the length of the face.
   e. The top of the ears is level with the eyebrows.
   f. The length of the ear from top to bottom is one-third the length of the face.

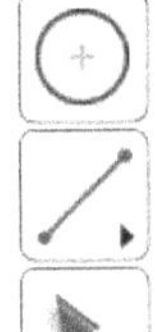

Use the **Compass** tool for the head (and for eyes if you want).

Use the **Segment** tool for parts of the face.

Use the **Arrow** tool to select and drag objects. Use a selection rectangle when needed.

Choose **Edit | Undo** to back up if you need to.

Select objects and choose **Display | Color** to change their color.

Use the **Text** tool to sign your name.

EXPLORE MORE

3. Go to page "Funny Face." Draw a face that does *not* follow the guidelines. Record the proportions of the face.

Jump Along: Positive and Negative Integers on the Number Line

INTRODUCE

Project the sketch for viewing by the class. Expect to spend about 15 minutes.

1. Open **Jump Along Integers.gsp.** Go to page "Reach Carrot." The rabbit should be located at 0 and the carrot at a positive integer no greater than 10. Tell students, ***Today you'll give directions to a rabbit on this number line so that it jumps to the carrot. Where is the rabbit and where is the carrot?*** Take responses.

2. ***There are two ways to control the rabbit's jumps. You can change how many jumps the rabbit will take, and you can change the size of the jumps. Where do you predict the rabbit will land when I press* Jump Along*?*** Take responses and then press the button. Discuss the trace the rabbit has left of its jumps.

Although it is possible to enter a non-integer value for the *Jump By* or *Number of Jumps* parameter, students will use only integer values in this activity.

3. If the rabbit has not reached the carrot, ask students how many more jumps are needed so that the rabbit will jump the rest of the way. To change the number of jumps, double-click the parameter *Number of Jumps* with the **Arrow** tool, enter a new value, and press OK. Then press *Jump Along* to watch the rabbit jump.

4. Press *Move Rabbit to 0* and *Erase Traces* to return the rabbit to 0. Now press *Move the Carrot* to have the computer choose a new, random location for the carrot between −10 and 10. Press the button several times, if necessary, until the carrot moves to a negative number. For our purposes here, we'll assume the carrot is at −7.

5. ***Where is the carrot?*** If this is students' introduction to negative numbers, have them observe that the number line has numbers less than 0. Explain that these are called *negative numbers.* Discuss the sign before the 7 and clarify that it is not read as *minus.*

   You may also wish to introduce the term *integer.* An integer is any counting number such as 0, 1, 2, and 3, as well as negative numbers such as −1, −2, and −3. The rabbit and the carrot will always be at integer locations on the number line in this activity.

Note that although *Number of Jumps* = −7 and *Jump By* = 1 describes a jump that should get to −7, the *Number of Jumps* parameter must always be greater than 0 for this activity.

6. ***How can I get the rabbit to the carrot?*** Invite discussion and try students' suggestions. For any trial, you can move the rabbit back to 0 and erase the traces using the action buttons. One way students may think about the question is this: If a *Jump By* value of 1 made the rabbit jump a distance of 1 to the right, perhaps a *Jump By* value of −1 will make the rabbit jump 1 to the left. In fact, when the rabbit takes 7 jumps of size −1 it lands at −7.

7. ***Now you're going to use the model on your own. But the rabbit won't start at 0 each time. It will start wherever it currently is. Right now, the rabbit is going to start at −7.*** Press *Move the Carrot.* ***How can the rabbit jump to the carrot?*** Solve this problem and one or two more with the class. Introduce the two rules for moving the rabbit: The value of *Number of Jumps* must be greater than 0, and the value of *Jump By* must be 1 or −1. If you haven't already, model pressing *Erase Traces* and tell students they can use this action button whenever they want.
8. Distribute the worksheet. Direct students' attention to the table. For each new location of the carrot, students will record these four pieces of information: where the rabbit starts, the location of the carrot, the value of *Number of Jumps* and the value of *Jump By.* Point out that the starting location of the rabbit will always be the same as the location of the carrot in the previous row of the table.

DEVELOP

Expect students at computers to spend about 30 minutes.

9. Assign students to computers and tell them where to locate **Jump Along Integers.gsp.** Encourage students to ask a neighbor for help if they have questions about using Sketchpad.
10. Let pairs work at their own pace. Here are some pointers related to the sketch to be aware of as you circulate.
    - Remind students to keep the value of the *Number of Jumps* parameter greater than 0 and the value of the *Jump By* parameter at 1 or −1.
    - If students enter numbers that don't land the rabbit on the carrot, they can move the rabbit back to where it began and try again. To do this, they simply drag the point that sits below the rabbit. Alternatively, if they don't reach the carrot, they can keep the rabbit where it is and figure out how it can jump again to reach the carrot.
    - When students no longer want to see existing traces, they should press *Erase Traces.*
    - Because the location of the carrot is picked randomly by the computer, it might happen that the carrot doesn't move when *Move the Carrot* is pressed. If so, students should press the button again. Also, you may wish to encourage students to press more than once in order to set up problems they think are different from the kinds they have solved so far.

11. As you circulate, listen to students' conversations. Here are some questions to ask and things to notice.

    - Notice students' vocabulary. Are they using the terms *negative* and *positive* as they describe the locations of the rabbit and carrot? Ask them to tell you the locations.
    - ***How are you deciding whether to make* Jump By *1 or −1?*** Students will likely notice that when *Jump By* is 1, the rabbit faces and jumps to the right. When *Jump By* is −1, the rabbit faces and jumps to the left. When the rabbit faces right, it is counting up from where it started. When it faces left, it is counting back from where it started.
    - Look for examples in which the rabbit and carrot are at locations with opposite signs. For example, suppose the rabbit is at −4 and the carrot is at 3. Ask students how they figured out the number of jumps the rabbit needed to make. One helpful way to think about such problems is to look at the distance from 0 to the rabbit and from 0 to the carrot. In this example, the rabbit is 4 units from 0 and the carrot is 3 units from 0 in the opposite direction. So the carrot is 4 units plus 3 units—7 units in all—away from the rabbit.

12. Encourage students who finish early to do the Explore More. The challenge is to find numbers for *Jump By* other than 1 or −1 that allow the rabbit to jump to the carrot. Students might also look for more than one way the rabbit can jump to reach the carrot. If the rabbit is at −2 and the carrot is at 4, for example, the rabbit can jump by 2's or 3's to reach the carrot. If the rabbit is at −2 and the carrot is at 5, the rabbit can take one jump of 7 to reach the carrot.

SUMMARIZE

Project the sketch. Expect to spend about 15 minutes.

13. Gather the class. Students should have their worksheets with them.

    Go to page "Make Your Own" and make sure the rabbit is at 10 and *Jump By* is equal to 1. Ask students how the rabbit can reach the carrot. Following their directions, change *Number of Jumps* to 10, and have the rabbit jump to the carrot.

    Press *Move Rabbit to 0,* drag the carrot to −10, and explain that you will keep *Number of Jumps* equal to 10. Again, ask students how the rabbit can reach the carrot. Have students tell you to change *Jump By* to −1. Press *Jump Along.*

14. ***What is the same and what is different about jumping from 0 to 10 and from 0 to −10?*** Here are sample student responses.

    When the rabbit jumps to 10, it faces to the right. When the rabbit jumps to −10, it faces to the left.

    When the rabbit jumps by a positive number, it moves in one direction. When the rabbit jumps by a negative number, it moves in the other direction.

    When the rabbit jumps by a positive number, it's the same as counting up. When the rabbit jumps by a negative number, it's the same as counting back.

    The rabbit jumped the same distance to go to 10 as it did to go to −10. That's because 10 and −10 are the same distance from 0, only on opposite sides of it.

    The jumps that go from 0 to −10 look like a reflection of the jumps that go from 0 to 10.

15. Focus the discussion on determining the distance and direction from one integer to another and on using the number line model to understand the order of positive and negative integers.

    Change *Jump By* to 1. ***Suppose the rabbit can make 5 jumps of 1 to get to the carrot. Where could the rabbit and carrot be?*** Have students talk in pairs or small groups and then take responses. Discussion should bring out the three possibilities and one big idea:

    - The rabbit and carrot could both be at locations greater than 0, such as 3 and 8.
    - The rabbit and carrot could both be at locations less than 0, such as −6 and −1.
    - The rabbit could be at a location less than 0, such as −2, and the carrot could be at a location greater than 0, such as 3.
    - Develop the idea that as long as the carrot is 5 units to the right of the rabbit, it is possible for their locations to be either positive or negative.

    Change *Jump By* to −1. ***Now, suppose the rabbit can make 5 jumps of −1 to get to the carrot. Where could the rabbit and carrot be?*** Provide time for discussion. Elicit the idea that, now, as long as the carrot is 5 units to the left of the rabbit, it is possible for their locations to be either positive or negative.

EXTEND

1. ***What other questions can you ask about jumping along a number line that shows both positive and negative numbers?*** Encourage student curiosity. Here are some sample student queries.

   Just by knowing the location of the rabbit and the location of the carrot, is it possible to tell whether the rabbit can reach the carrot by jumping by 2's?

   Suppose I tell you where the rabbit landed, the number of jumps, and the size of each jump. Can you figure out where the rabbit started?

   Suppose the rabbit jumps part way to the carrot, stops, and then jumps the rest of the way using a different jump size. How does that change things?

2. Have students make jump-along problems for others to solve. Page "Make Your Own" of the model allows students to drag the carrot and rabbit to positions of their choosing. As well, by dragging the point that sits at 1, students can change the scaling of the number line to include numbers beyond the -10 to 10 range.

Jump Along Integers

Name:

Help a rabbit reach a carrot on the number line.

1. Open **Jump Along Integers.gsp.** Go to page "Reach Carrot."

   There are two rules. The rabbit must jump by 1 or −1. The number of jumps must be greater than 0.

   To change the value of *Number of Jumps* or *Jump By,* select the number and use the + or − key on the keyboard.

   Press *Jump Along* to make the rabbit jump.

   Press *Move the Carrot* to make a new problem.

2. Record your work in the table.

Rabbit Starts At	Location of Carrot	Number of Jumps	Jump By

EXPLORE MORE

3. Continue to move the carrot. Try to reach the carrot using *Jump By* numbers other than 1 or −1.

   Record your work on the back of this sheet.

Right or Left: Adding and Subtracting Integers

ACTIVITY NOTES

INTRODUCE

Project the sketch for viewing by the class. Expect to spend about 5 minutes.

1. Open **Right or Left.gsp** and go to page "Addition." Enlarge the document window so it fills most of the screen.
2. Explain, ***Today you're going to use Sketchpad to add and subtract integers on a number line.*** You might remind students that integers are the set of whole numbers and their opposites, and ask them to give real-world examples of integers.
3. ***First I'll show you how the Sketchpad model works.*** Model worksheet steps 1–5. Here are some tips.
   - Introduce the addition problem in the sketch. ***What addition problem is shown?*** [8 + 5] ***Are the numbers in the problem positive or negative?*** [Both positive] ***Will the answer be to the right or to the left of the first number?*** [Right] ***Will it be to the right or left of zero on the number line?*** [Right] Ask a few volunteers to share their predictions. Then press *Add.* Press *Reset,* and then press *Add* a second time.

In this model each integer is represented by a vector. The term *arrow* is used because students may not be familiar with the term *vector* yet. A vector has both magnitude and direction, so practice with this model will help students understand how the signs of the addends come into play.

   - ***What does the first arrow represent?*** [8] ***What does the second arrow represent?*** [5] ***Why do both arrows point to the right?*** [Adding two positive numbers] ***How can you find the answer to 8 + 5 using the model?*** Students may make a response such as the following: *The second arrow starts from the end of the first arrow, so where the second arrow ends is the answer. The second arrow ends at 13.*
   - Now press *Reset,* and drag the circles to model 2 + (−6). ***By dragging the circles, I can change the problem. What is the problem now?*** [2 + (−6)] ***Why does the second arrow point left?*** [It represents a negative number.] ***Why does an arrow that represents a negative number point left?*** [Negative numbers are to the left of zero on a number line] ***How many units long is each arrow?*** [The arrow representing 2 is 2 units long; the arrow representing −6 is 6 units long.] ***Predict whether the answer will be to the right or to the left of zero.***
   - ***If I press* Show Steps*, we can see the solution modeled in steps.*** Press *Show Steps,* and then press each numbered button to see the model step by step. ***What is the answer to 2 + (−6)? How do you know using the model?*** Here is a sample student response: *The answer is −4*

because that is where the second arrow ended up after placing it at the end of the first arrow.

- You might consider doing another sample problem without pressing *Reset.* Students may gain more insight from leaving the arrows visible on the number line as they drag the circles to change the problem.

4. Ask students to consider the relationship between addition and subtraction of integers as they work. ***As you are adding and subtracting integers using the Sketchpad model, think about how adding and subtracting integers are related. How are they similar? How are they different?***

DEVELOP

Expect students at computers to spend about 35 minutes.

5. Assign students to computers and tell them where to locate **Right or Left.gsp.** Distribute the worksheet. Tell students to work through step 25 and do the Explore More if they have time. Encourage students to ask their neighbors for help if they are having difficulty with Sketchpad.

6. Let pairs work at their own pace. As you circulate, here are some things to notice.

   - In worksheet step 3 and for all activity questions, encourage students to write clear and detailed explanations using complete sentences. By clearly describing what they observe, students acquire a strong mental image of operations with integers. If time is limited, you might have students write their explanations for homework.

   - In worksheet step 8, have students predict what will happen in the Sketchpad model before pressing any buttons. ***What will the model of*** $-6 + (-3)$ ***look like? Why?*** Try to get students to concentrate on the behavior of the model rather than on the numeric answer.

   - In worksheet steps 10 and 11, encourage students to explore these questions by dragging to change the values of the integers without pressing *Reset.* Students can quickly view several problems before making a conjecture.

   - In worksheet step 12, students must interpret different parts of the Sketchpad model. As you walk around, observe students to be sure they understand each part and can model any problem they are given. When students successfully model all the problems, ask them to look for patterns. Students may notice that the sign of the answer is the same as the sign of the longer arrow. They may not recognize this as the integer with the greater absolute value; that's okay. Students are focusing on the visual model at this time. Discuss absolute value later.

- In worksheet step 14, students start subtracting integers. The concept of additive inverse is not explicitly named, but it plays a prominent role in the model. Ask students to think about why the second integer is flipped in a subtraction problem.
- As students are creating new subtraction problems, ask them to predict what the model will do each time before pressing the action buttons. ***Can you predict what the model of this subtraction problem will look like? Why do you think it will act that way?*** Students can test their conjectures using the step-by-step buttons.
- In worksheet step 24, ask students what patterns they see and how they could predict the answer from the two numbers being subtracted.
- In worksheet step 25, students are asked to write an addition problem using the same first number and the same answer. Students can test their addition problems by going to page "Addition." Switching back and forth between the two pages will reinforce the idea of using addition to rewrite a subtraction problem: To find the answer to a subtraction problem, you add the additive inverse (opposite) of the second number.
- If students have time for the Explore More, they will investigate the behavior of addition and subtraction independent of specific values, and they will use special cases to identify the position of zero on the number line.

SUMMARIZE

Project the sketch. Expect to spend about 20 minutes.

7. Gather the class. Students should have their worksheets with them. Begin the discussion by opening **Right or Left.gsp** and going to page "Addition." Work through the different types of addition problems with the class.
   - Have volunteers model the problems they recorded for worksheet step 9. ***What happens in the model when you add two negative integers?*** Students may make this sample response: *When adding two negative integers, the arrows both point left, so the answer is always*

negative. ***How does this compare to adding two positive integers?*** Students may reply that in both cases the arrows point in the same direction. With positive integers, the arrows point right. With negative integers, the arrows point left.

- Next have volunteers model the problems in worksheet step 12. ***What happens in the model when you add a positive and a negative integer?*** Students may make the following response: *If the negative integer is greater, the arrow pointing left will be longer, so the answer will be negative. If the positive integer is greater, the arrow pointing right will be longer, so the answer will be positive.*
- At this point, you may wish to introduce the term *absolute value* and the absolute value symbol. **Absolute value** ***is the distance a number is from zero. What represents the absolute value of a number in this model?*** Help students see that the length of an arrow is the distance from zero. ***What is the absolute value of −2?*** [2] ***What is the absolute value of 2?*** [2] Work through several problems with the class, each time focusing on the length of the arrow. Students should understand that opposites, or additive inverses, have the same absolute value. ***Can the absolute value of a number ever be negative?*** Students should realize that because distance is a positive value, the absolute value can never be negative.
- ***When adding a positive and a negative integer, how can you look at the numbers and tell whether the answer will be positive or negative?*** Students may make the following responses.

  The sign of the number with the longer arrow will be the sign of the answer.

  The sign of the number with the greater absolute value will be the sign of the sum.

8. Go to page "Subtraction." Have volunteers model subtracting two positive integers, a negative and a positive integer, a positive and a negative integer, and two negative integers. ***How are adding and subtracting integers related? How are they similar? How are they different?*** Students may respond with the following answers.

   When you subtract two integers, you flip the second number, so its arrow points the other way. You don't do that with addition.

In subtraction, after you flip the second number, the model is similar to addition. The answer is where the second arrow ends.

Subtraction is just adding the second number flipped.

In subtraction you are adding the opposite of the second number.

9. If time permits, discuss the Explore More. Have students explain how they determined the position of zero.
10. ***Explain the different ways you can get a negative answer when you subtract two integers.*** You may wish to have students respond individually in writing to this prompt. Here are the possible ways: If both integers are positive, the second integer must be greater than the first one. If the first integer is negative and the second integer is positive, the difference will be negative. If both integers are negative, the second integer's absolute value must be smaller than that of the first integer.

EXTEND

1. ***When you add two integers, does the order matter? In other words, is*** $-3 + 5$ ***the same as*** $5 + (-3)$***?*** Have students use the sketch to explain their answers. The order does not matter when you add two integers. The arrows determine how far you go and in which direction. It doesn't matter whether you follow the first arrow and then the second or whether you follow the second arrow and then the first.
2. ***When you subtract two integers, does the order matter? In other words, is*** $-3 - (-5)$ ***the same as*** $-5 - (-3)$***? Explain in terms of the model why your answer makes sense.*** The order does matter when you subtract integers, because only the second arrow is flipped. More sophisticated students will observe that the order matters only if the second integer is nonzero, because flipping zero has no effect.
3. ***What questions occurred to you while you were adding and subtracting integers?*** Encourage curiosity. Here are some sample student queries.

   Why do you flip the second arrow when subtracting?

   If you subtract a positive number from any number, is the result always to the left of the first number?

   Can this model work for multiplying and dividing integers?

ANSWERS

3. In their final positions, the second arrow starts from where the first arrow ends, and the answer (13) is at the end of the second arrow.
6. Answers will vary. Problems should include only positive integers.
7. Each bottom arrow is exactly the same size and direction as the corresponding top arrow.
8. The sum of $-6 + (-3)$ is -9.
9. Answers will vary. Problems should include only negative integers.
10. Whether adding two negative or two positive integers, both arrows go the same way, taking the sum farther away from the center of the number line (farther from zero). The difference is that the arrows go to the right when the numbers are positive and to the left when they are negative.
11. When you add two negative integers, you cannot get a positive sum. Both numbers take the sum in the negative direction from zero, so the sum must be negative.
12. $7 + (-4) = 3$ $\quad$ $-4 + 7 = 3$

    $-6 + 2 = -4$ $\quad$ $2 + (-6) = -4$

    $-3 + 7 = 4$ $\quad$ $3 + (-7) = -4$

    $2 + (-5) = -3$ $\quad$ $-2 + 5 = 3$
13. When you add a positive and a negative integer, the integer that has the greater absolute value tells you whether the answer will be positive or negative. In other words, the sign of the answer is the same as the sign of the longer arrow.
16. During the animation, the arrow for 5 flips from the right to the left. This shows which way the second arrow must go in order to subtract it from the first.
18. Answers will vary. This step creates the additive inverse by flipping the arrow.
19. Problems will vary. Problems should include those in which the second integer (*subtrahend*) is greater than the first integer (*minuend*).
20. If both integers are positive, the answer will be positive if the first integer is greater, and it will be negative if the second integer is greater.

21. Some students will write direct observations, and others will interpret those observations. Typical answers will be similar to these.

    Observation: In this problem, $4 - (-3)$, the second arrow starts out pointing to the left, so when it flips, it turns around and points to the right.

    Interpretation: The second number starts out negative, so when it flips, it becomes positive.

22. Problems will vary. The first integer is positive and the second integer is negative, so after flipping, both arrows point to the right. The answer must be positive.

23. Problems will vary. The first integer is negative and the second integer is positive, so after flipping, both arrows point to the left. The answer must be negative.

24. $7 - (-4) = 11$ $\quad -4 - 7 = -11$

    $-3 - 8 = -11$ $\quad -3 - (-8) = 5$

    $2 - (-7) = 9$ $\quad -2 - 7 = -9$

    $-6 - (-5) = -1$ $\quad -5 - (-6) = 1$

25. $7 + 4 = 11$ $\quad -4 + (-7) = -11$

    $-3 + (-8) = -11$ $\quad -3 + 8 = 5$

    $2 + 7 = 9$ $\quad -2 + (-7) = -9$

    $-6 + 5 = -1$ $\quad -5 + 6 = 1$

    Subtracting is the same as adding the opposite.

26. For addition, the sum moves in the same direction and at the same speed as the movement of either integer. For subtraction, the difference moves in the same direction and at the same speed as the movement of the minuend (a), but it moves in the opposite direction as the movement of the subtrahend (b).

27. For addition, move one integer until the sum is equal to the other integer. For subtraction, move the two integers to the same location so the difference is at zero, or move the subtrahend (b) until the difference equals the minuend (a).

Right or Left?

Name:

In this activity you'll add and subtract integers on a number line.

EXPLORE

1. Open **Right or Left.gsp** and go to page "Addition." If necessary, drag the circles to model the addition problem 8 + 5.

2. Press *Add.* Observe the model in action.

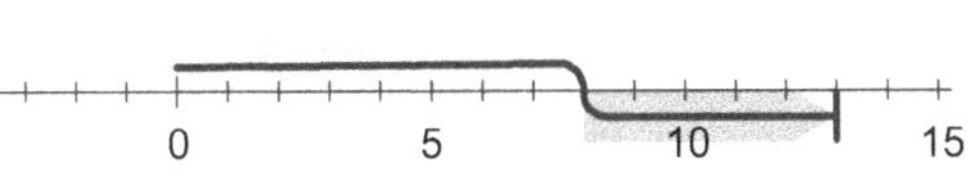

3. How does the final position of the arrows on the number line show the answer for this addition problem?

4. Press *Reset.* Drag the circles to model 2 + 6.

5. This time, press *Show Steps.* Then press each numbered button in order to see the model step by step.

6. Drag the circles to model another addition problem using only positive integers. Record your problem and its result.

7. How do the two top arrows in the sketch relate to the two bottom arrows?

8. Model −6 + (−3). What is the sum?

drag −6

drag + −3

9. Model two more addition problems using negative integers. Record each problem and its result.

10. How is adding two negative integers similar to adding two positive integers? How is it different?

11. Can you add two negative integers and get a positive sum? Explain.

12. Model the following eight addition problems. Record each problem and its answer.

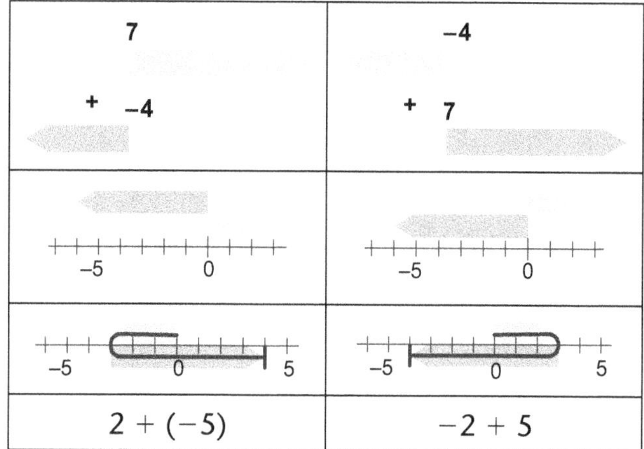

13. When you add a positive and a negative integer, how can you look at the numbers and tell whether the answer will be positive or negative?

Now you'll explore a subtraction model.

14. Go to page "Subtraction." If necessary, drag the circles to model the subtraction problem 8 − 5.

15. Press *Subtract.* Observe the model in action.

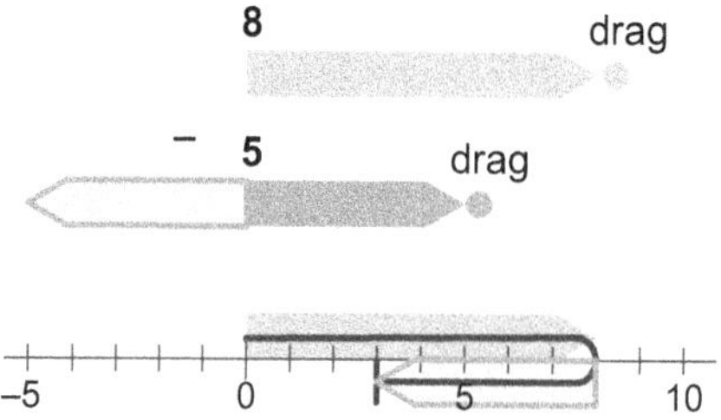

16. During the animation, what happens to the arrow for the integer 5? What does this show?

17. Press the *Reset* button. Then drag the circles to model 2 − 6.

18. This time, show the animation step by step. Describe in your own words what the step *3. Make Inverse* does.

19. Drag the circles to model two more subtraction problems that use positive integers but have a negative result. Record each problem and its result.

20. If both integers in a subtraction problem are positive, how can you tell whether the answer will be positive or negative?

21. Model 4 − (−3). What's different about the step *3. Make Inverse* this time?

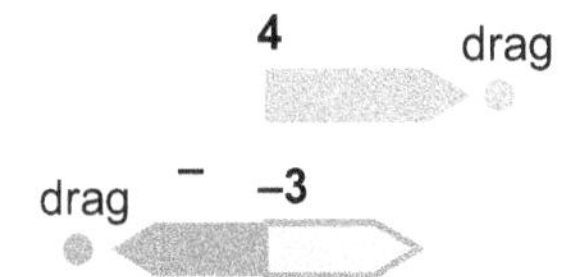

22. Model two more problems in which the first integer is positive and the second integer is negative. Record each problem. What do these models have in common?

23. Model three problems in which the first integer is negative and the second integer is positive. Record each problem. What do these models have in common?

24. Model the following eight subtraction problems. Record each problem and its answer.

7 − −4	−4 − 7
−10 −5 0	0 5
$2 - (-7)$	$-2 - 7$
$-6 - (-5)$	$-5 - (-6)$

25. For each subtraction problem in step 24, write an addition problem that has the same first integer and the same answer. What do you notice?

EXPLORE MORE

26. Go to page "Explore More." You will see two number lines, one that shows the sum $a + b$ and another that shows the difference $a - b$. With the numbers hidden, drag a and b and observe the behavior. How do addition and subtraction behave similarly? How do they behave differently?

27. On each number line, use your number sense to figure out where zero must be located. Drag a gold arrow to mark this location, and then press *Show Numbers* to check your answer.

Jeff's Garden: Area Model of Fraction Multiplication

ACTIVITY NOTES

INTRODUCE

Project the sketch on a large-screen display for viewing by the class. Expect this part of the activity to take about 40 minutes.

1. Before using the sketch, pose the problem below. As you do, record the following for students to reference.

   $\frac{3}{5}$ of the garden is Jeff's

   $\frac{2}{3}$ of his part is for pumpkins

   Jeff earns extra money by selling produce he grows in his grandmother's garden. This year, his grandmother will allow him to use $\frac{3}{5}$ of her whole garden. Jeff has decided that he will use $\frac{2}{3}$ of his part to grow pumpkins. What amount of his grandmother's garden will he use to grow pumpkins?

Some students may have learned an algorithm for multiplying fractions. Asking these students to make a drawing ensures that they think about making sense of the problem.

2. Provide paper and explain that students should use a drawing to show this situation. ***When you come up with an answer, make sure you can explain why it makes sense.*** Allow students to grapple with this, working in pairs or groups. As you circulate, ask questions to help students persist in reasoning about the problem. Because students often don't relate the word "of" to multiplying, don't expect them to think in terms of multiplying the fractions at this point.

   What does it mean to have $\frac{3}{5}$ of a whole?

   What could you show next in your drawing that would help you think about the problem?

   Do you think Jeff is using more than, less than, or exactly half of the whole garden for pumpkin growing? Does your drawing make sense, given that idea?

3. Lead a discussion of students' drawings. Invite students' questions as well as their attempted representations. There is more than one way to draw this situation. Give students time to consider any different drawings that seem correct.

Becoming Familiar with the Model

4. Open **Jeff's Garden.gsp.** Go to page "Area Model." ***Let's see how we can use this Sketchpad model to represent the garden problem.*** Follow these steps.

To change a numerator or denominator, double-click the number. In the dialog box that appears, enter a new value.

- ***The rectangle represents grandmother's garden, the whole garden.***
- ***To represent Jeff's part, we need to show $\frac{3}{5}$ of the whole garden.*** Change the denominator of the second fraction to 5. Students will see that the rectangle is now divided into fifths by vertical lines.

  Change the numerator to 3. Three of the fifths are now colored. Restate that this is Jeff's part of the garden, $\frac{3}{5}$ of the whole garden.

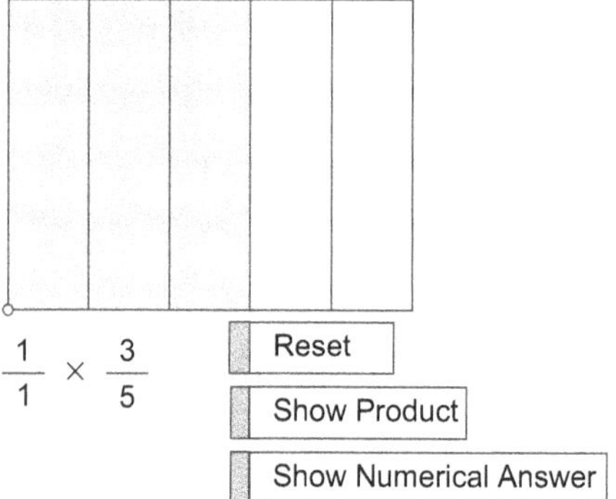

- ***Now we want to show $\frac{2}{3}$ of Jeff's part. What would his part of the garden look like divided into thirds? Picture this in your mind.*** Set the denominator of the first fraction to 3, and then drag the point *across the colored fifths only.* Students will see that two horizontal lines divide the colored area into thirds.

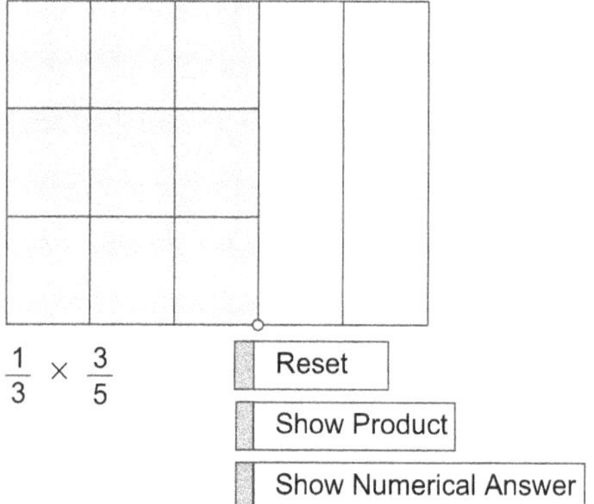

- ***How many thirds of his garden will Jeff plant in pumpkins?*** [Two] ***Let's show $\frac{2}{3}$ of Jeff's part.*** Change the numerator to 2 and press *Show Product.* Two of the horizontal regions are now colored. One of the thirds (the part not in pumpkins) is not.

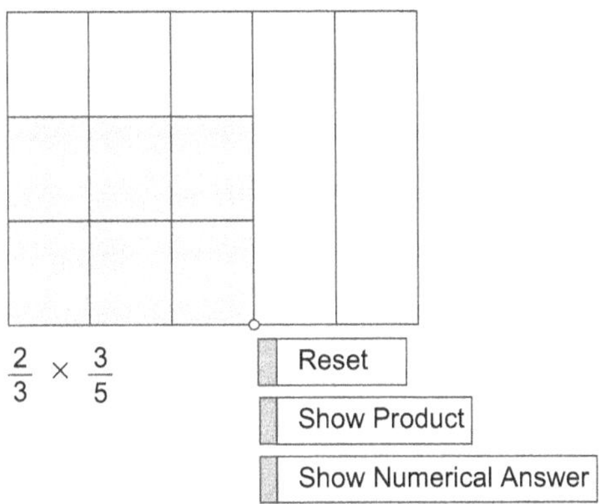

- ***What part of grandmother's garden—the whole garden—will Jeff plant in pumpkins?*** Give students time to think, and then drag the point the remainder of the way across the rectangle.

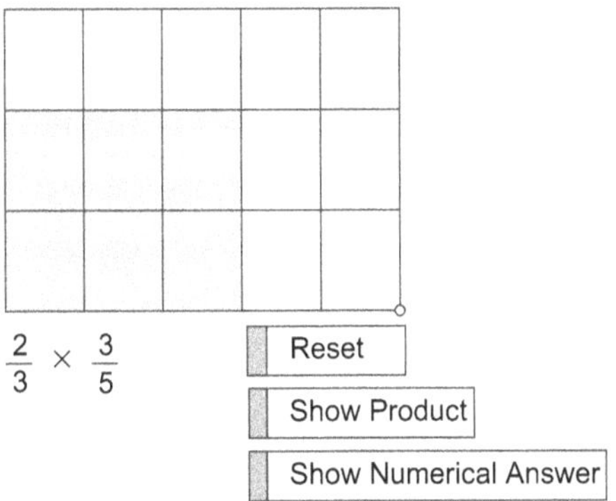

Ask again, ***What part of grandmother's whole garden will be planted in pumpkins?*** The discussion should yield these ideas: Grandmother's garden is now divided into 15 equal parts; six of the parts (the colored region) are the amount Jeff will plant in pumpkins; so, Jeff will use $\frac{6}{15}$ of his grandmother's garden for pumpkins.

By asking whether the answer makes sense, you are modeling a question you want students to be asking themselves as they carry out computations.

- ***Does this answer make sense?*** Solicit thinking such as this student sample: *You can see that the garden is divided into fifteenths. Jeff's garden is 9 parts of the whole and 6 of those parts are for pumpkins. I know 6 is $\frac{2}{3}$ of 9, so $\frac{6}{15}$ makes sense for the $\frac{2}{3}$ of Jeff's garden.*
- Press *Show Numerical Answer* for confirmation of the answer.

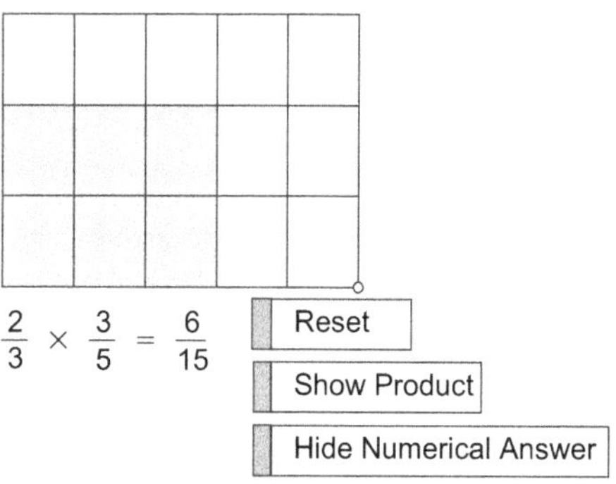

5. Provide a moment to reflect. ***How does this model compare to the way you thought about the problem and the drawing you made at the start of this activity?*** Take some responses, or have students discuss in pairs or small groups.

6. Model using the *Reset* button. Pressing the button removes the horizontal lines, the shaded area for the product, and the numerical answer. Students will need to set the fractions that are multiplied back to $\frac{1}{1}$ by changing the numerators and denominators themselves. Doing so will display an undivided rectangle.

DEVELOP

Expect students working at computers to take about 30 minutes.

7. Assign students to computers and tell them where to find **Jeff's Garden.gsp.** Distribute the worksheet. Explain that students should work on steps 1 and 2. ***Using the model, make sense of the problems. Record your answers. Do the Explore More if you have time.***

8. As you circulate, listen for ways students are making sense of fraction multiplication. Here are some things to notice.

   - If you need to help some students to reason as they work on Year 1 in the table, pose questions such as these: ***Do you need to start with $\frac{3}{7}$? Suppose you started by thinking about $\frac{3}{7}$? What would $\frac{1}{3}$ of $\frac{1}{7}$ look like?***

   The important thing here is that students not use the model, without thinking, merely following the steps they have learned.

   - Notice any students who are making conjectures about other ways of solving fraction multiplication problems. If some students posit that the numerator in the product can be found by multiplying the numerators of the factors (and the same for the denominator in the product), ask, ***Can you explain why that should work? Do you think it will work for every problem of this type?*** Don't confirm for students at this point that this method will always work.

Be on the lookout for any pairs who are using the model in an automatic way without making sense of each action they carry out. Ask, *What are you showing now? What part of the problem does that represent?*

- Some pairs may devise shortcuts for using the model. If you observe shortcuts being used, ask students to explain what they are doing and how they are thinking. If students are making sense of what they are doing, they should continue. For example, students may change *both* numerators and denominators to match the values in a problem before they drag the point and press *Show Product.* Or students may manipulate the model as demonstrated, but omit the last step of extending the horizontal lines across the whole garden. Most likely, they are able to mentally extend the lines and determine the pieces the whole rectangle is divided into.

SUMMARIZE

Project the sketch for viewing by the class. Expect to spend about 30 minutes.

9. Students should have their worksheets. Lead a discussion to explore the idea that multiplication by a fraction less than one results in a product smaller than the number being multiplied. (This is true for multiplying a whole number by a fraction and for multiplying a fraction by a fraction.) Begin by asking the following questions.

   You started with $\frac{3}{5}$ and multiplied it by $\frac{2}{3}$. Was the product (the amount planted in pumpkins) bigger or smaller than the number you started with, $\frac{3}{5}$?

   You multiplied and got a product smaller than the number you started with. Does that make sense?

   From their experiences with whole numbers, students assume that multiplication results in a product larger than the factors. Provide time for the class to grapple with this in order to develop a conceptual foundation that makes sense of fraction multiplication. Invite students to model at the computer as the class considers this issue.

   The discussion should bring out the idea that multiplying by a fraction involves taking *a part of* what you started with. Multiplying a fraction by a fraction involves *taking a part of a part.* For example, $\frac{1}{2} \times \frac{1}{5}$ means taking $\frac{1}{2}$ *of* $\frac{1}{5}$. You end up with less than you started with.

It is important to spend time developing the idea that multiplying a fraction by a fraction is finding a fraction *of* a fraction.

10. Discuss the problem in worksheet step 2. Note whether any students have related this problem to Jeff's garden in Year 2 (in worksheet step 1). In Year 2, Jeff plants $\frac{4}{5}$ of half the garden in pumpkins. In step 2, he plants half of $\frac{4}{5}$ of the garden in pumpkins. The answer is the

same in both cases, $\frac{2}{5}$. This suggests that multiplication of fractions is commutative. If students don't propose this idea, introduce it. Provide time for the class to explore one or two other problem pairs that are easy to visualize. ***What is*** $\frac{1}{2}$ ***of*** $\frac{1}{4}$***? What is*** $\frac{1}{4}$ ***of*** $\frac{1}{2}$***?***

11. Present the following problem, which gives large values for the denominators. ***Another year, Jeff's grandmother was not well. She gave him*** $\frac{7}{8}$ ***of the garden and kept a small strip for herself. Jeff planted*** $\frac{3}{15}$ ***of his area in green beans. What part of the whole garden did he plant in beans?***

The class is likely to anticipate that the model will display more parts in the solution than students wish to count. The intention here is to prompt them to look for strategies other than counting all the parts, if they haven't already. Provide a few minutes for students to work on the problem, recording on the back of their worksheets.

Ask students to share how they solved the problem. You may wish to say, ***I saw some of you using shortcuts.*** Call on students whom you noted using shortcuts, or ask for volunteers to model at the computer.

Some students are likely to have noticed that for all the garden problems, the product can be found by multiplying the numerators and multiplying the denominators of the factors. If this isn't suggested, ask, ***How are the numerators and denominators represented in the model?*** Set the model for $\frac{2}{3} \times \frac{3}{5}$, the problem you first modeled. Provide ample time for students to compare the symbolic representation with the area model.

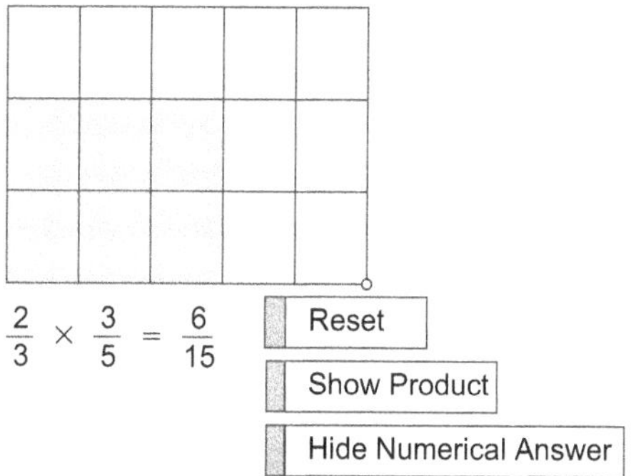

Invite explanations such as these student examples.

The colored area shows the numerators, $2 \times 3 = 6$, *and the whole rectangle shows the denominators,* $3 \times 5 = 15$.

Problems that come at the mathematics from a different angle are good for both strengthening and assessing student understanding.

The numerators are represented by two rows of three, and the denominators are represented by the three rows of five.

12. Discuss the Explore More problem, worksheet step 3. This is a working backward problem. Students know the part of the entire garden planted in pumpkins by Jeff and are asked to determine the part of Jeff's area that is planted in pumpkins.

EXTEND

1. Present problems with missing factors. Have students use the Sketchpad model to determine and/or check their answers.
2. Develop in students the habit of using computational estimation to anticipate and check computations for reasonableness. Present computations like these, and ask students to estimate whether the product in each computation will be greater or less than 1/2. Ask students to explain their reasoning.

   $\frac{5}{6} \times \frac{1}{3}$ $\quad$ $\frac{4}{5} \times \frac{7}{8}$ $\quad$ $4 \times \frac{8}{9}$

3. Have students write another problem like the Explore More. Students should exchange problems and solve. You might use this as an individual assessment.

ANSWERS

1.

Year 1	$\frac{3}{21}$
Year 2	$\frac{4}{10}$
Year 3	$\frac{21}{40}$
Year 4	$\frac{20}{72}$
Year 5	$\frac{10}{18}$

2. $\frac{4}{10}$, or $\frac{2}{5}$

3. $\frac{1}{2}$

Jeff's Garden

Name:

1. For the next five years, Jeff's grandmother gives him part of her garden. Jeff uses part of his area to grow pumpkins. What part of the whole garden does Jeff plant in pumpkins each year? Use **Jeff's Garden.gsp** to help you complete the table.

	Part of the Whole Garden Jeff Gets	**Part of Jeff's Area He Plants in Pumpkins**	**Part of the Whole Garden Jeff Plants in Pumpkins**
Year 1	$\frac{3}{7}$	$\frac{1}{3}$	
Year 2	$\frac{1}{2}$	$\frac{4}{5}$	
Year 3	$\frac{7}{10}$	$\frac{3}{4}$	
Year 4	$\frac{4}{9}$	$\frac{5}{8}$	
Year 5	$\frac{2}{3}$	$\frac{5}{6}$	

2. Another year, Jeff's grandmother gave him $\frac{4}{5}$ of the garden. He planted $\frac{1}{2}$ of it in pumpkins. What part of the whole garden did he plant in pumpkins? __________

EXPLORE MORE

3. One year, Jeff was given $\frac{2}{3}$ of the garden. He planted a part in pumpkins. His grandmother said, "This year $\frac{1}{3}$ of the whole garden is planted in pumpkins." Grandmother did not plant pumpkins. What part of Jeff's garden was planted in pumpkins? __________

Feed the Mouse: Fraction and Decimal Relationships

ACTIVITY NOTES

INTRODUCE

Project the sketch for viewing by the class. Expect to spend about 5 minutes.

1. Open **Feed the Mouse.gsp.** Go to page "Mouse on Floor." Distribute the worksheet.

2. Explain, ***Today you'll use Sketchpad to feed a piece of cheese to a hungry mouse.*** Point out the square piece of cheese, the mouse, and the hole on the floor.

   Press *Feed the Mouse.* The cheese will drop and the mouse will move toward it. When the cheese lands, the mouse will take a bite. Press *Reset* to return the mouse and the cheese to their starting positions.

3. ***What's the width of the hole?*** $\left[\frac{1}{3}\right]$ ***What's the width of the cheese?*** $\left[\frac{1}{1}, \text{or } 1\right]$ ***That's a big piece of cheese for such a little mouse! Let's keep the width of the cheese less than 1. We need to be careful, though, that the width of the cheese isn't less than the width of the hole. If it is, the cheese will fall through the floor, and the poor mouse won't eat.***

DEVELOP

Continue to project the sketch. Expect to spend about 30 minutes.

Be sure students use the vocabulary *greater than, less than,* or *equal to* when describing the comparisons.

4. ***Let's keep the numerator of the cheese fraction at 1. What should we pick for the denominator?*** Take a suggestion. For this example, we'll assume that students propose a width of $\frac{1}{2}$ for the cheese.

   Model how to change the denominator of the fraction: Use the **Arrow** tool to double-click the denominator, and then enter 2 in the dialog box. Click OK. Press *Feed the Mouse.* The mouse will eat the cheese.

   Why does the mouse get fed? Students might explain that because $\frac{1}{2}$ is bigger than $\frac{1}{3}$, the cheese does not fall through the hole. ***What's another way to say that?*** [$\frac{1}{2}$ is *greater than* $\frac{1}{3}$ or $\frac{1}{3}$ is *less than* $\frac{1}{2}$.] Use the symbols $>$ and $<$ to show the comparisons on the board: $\frac{1}{2} > \frac{1}{3}$ and $\frac{1}{3} < \frac{1}{2}$.

   Students should record the fractions, the comparison, and whether the mouse was fed in the first row of the table on their worksheets.

5. Press *Reset.* Ask students to suggest a denominator for the cheese that will cause it to fall through the hole. Again, keep the numerator of the cheese at 1. Students might propose $\frac{1}{4}$. Change the cheese denominator to 4, and then press *Feed the Mouse.* The cheese sails through the hole and off the screen.

Why did the cheese fall through the hole? Students should respond that $\frac{1}{4}$ is less than $\frac{1}{3}$. Again, students should record the comparison on their worksheets.

6. Students commonly believe that a larger denominator means a larger fraction. With the denominator of the cheese fraction selected, repeatedly press the + key on your computer. As you do, the denominator of the cheese will keep increasing by 1 and the width of the cheese will decrease.

   Read the new width each time you press the + key to emphasize that a larger denominator means a smaller fraction. (You don't need to drop the cheese. The visual of the cheese getting smaller and smaller as the denominator gets larger and larger is sufficient.) The cheese becomes difficult, if not impossible, to see when it is smaller than $\frac{1}{40}$.

7. ***What do you think happens when the cheese and the hole are the same width?*** Some students may argue that the cheese will go through the hole; others will say that the cheese must be smaller to fit through the hole. Model what happens using $\frac{1}{3}$ and $\frac{1}{3}$. In this sketch, the cheese will not go through the hole when they are the same width. Write the two fractions on the board using the equal sign to show they are equivalent: $\frac{1}{3} = \frac{1}{3}$.

8. ***If we keep the width of the hole at $\frac{1}{3}$, what different fraction with the same denominator can we use for the cheese?*** Students may suggest $\frac{2}{3}$. Change the fraction and press *Feed the Mouse* to check. ***What if the width of the hole is $\frac{5}{8}$? What fraction with the same denominator can you choose for the width of the cheese so it doesn't go through the hole?*** Students may suggest $\frac{6}{8}$ or $\frac{7}{8}$. Students should realize that when fractions have the same denominator, students can compare just the numerators to see which is larger.

9. ***Now let's change the width of the hole to $\frac{1}{2}$. What size cheese should we try?*** Students' sense of benchmark fractions should help here. They may make suggestions such as $\frac{2}{3}$, $\frac{3}{4}$, $\frac{4}{5}$, or $\frac{9}{10}$—all fractions they know to be greater than $\frac{1}{2}$.

10. ***Let's change the width of the hole to $\frac{3}{4}$. Do you think $\frac{2}{3}$ will work for the width of the cheese?*** Take responses. Elicit or introduce the idea that another way to compare fractions is to look at their decimal equivalents. ***How do we write $\frac{3}{4}$ as a decimal?*** [0.75] ***How do we write $\frac{2}{3}$ as a decimal?*** [0.66] ***Which is larger?*** [0.75]

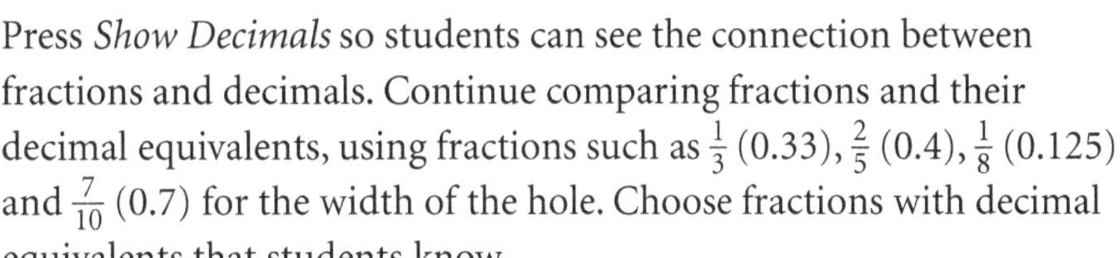

Press *Show Decimals* so students can see the connection between fractions and decimals. Continue comparing fractions and their decimal equivalents, using fractions such as $\frac{1}{3}$ (0.33), $\frac{2}{5}$ (0.4), $\frac{1}{8}$ (0.125) and $\frac{7}{10}$ (0.7) for the width of the hole. Choose fractions with decimal equivalents that students know.

11. Now change the width of the hole to $\frac{6}{7}$. The model will report that the decimal equivalent is 0.86. ***What's another way you can write 0.86 as a fraction?*** $\left[\frac{86}{100}\right]$ ***What ideas does that give you for the width of the cheese?*** Students may suggest $\frac{87}{100}$. ***Can you give me ten other fractions that will also work?*** Students should note that any fraction from $\frac{88}{100}$ up to $\frac{99}{100}$ can be the width of the cheese. ***Being able to work with both fractions and decimals is helpful.***

12. Some students may suggest using the "missing piece" strategy for working with any fraction whose numerator is 1 less than the denominator. Given a fraction like $\frac{6}{7}$, for example, a student might increase both the numerator and denominator by 1 and propose that $\frac{7}{8}$ will work for the cheese. This is true because $\frac{6}{7}$ is $\frac{1}{7}$ less than 1, whereas $\frac{7}{8}$ is $\frac{1}{8}$ less than 1. Because the missing piece $\frac{1}{7}$ is greater than the missing piece $\frac{1}{8}$, $\frac{6}{7}$ is less than $\frac{7}{8}$. Try other fractions whose numerators and denominators differ by 1, such as $\frac{8}{9}$, having students use either this strategy or the decimal approach to find a fraction for the cheese.

SUMMARIZE

Project the sketch. Expect to spend about 10 minutes.

13. Conclude the activity by asking students to summarize what they've learned. ***What strategies did you use to pick a size of cheese that fed the mouse?*** Students may make the following observations.

    I know that I can use a fraction with the same denominator as the width of the hole, but with a larger numerator.

    I know that I can use a fraction with the same numerator as the width of the hole, but with a smaller denominator.

    Sometimes I can use benchmarks. When the width of the hole was $\frac{1}{2}$, for example, I used a fraction that I knew was greater than $\frac{1}{2}$ for the cheese.

    If the width of the hole is not an easy fraction, I can convert it to a decimal and then write the decimal as a fraction using tenths or hundredths. Then I can easily name a fraction that is greater.

I can think of a fraction that has 1 more in the numerator and 1 more in the denominator. For example, I know $\frac{4}{5}$ is greater than $\frac{3}{4}$.

EXTEND

1. Give students an opportunity to work with the sketch at a later time. Have pairs create challenges for each other—one student providing the width of the hole and the other finding a width for the cheese that will feed the mouse.
2. Solve problems with improper fractions, picking widths of the hole that are greater than 1.
3. Go to page "Mouse in Cage." A mouse sits in a cage with two holes. The cheese must be small enough to fit through the hole in the top of the cage, but large enough not to fall through the floor of the cage.

   Students will need to compare three fractions. Work through problems with students in a similar way as before, having students order the fractions from smallest to largest. Note that depending on the size of the two holes, it may not be possible to feed the mouse at all!

Feed the Mouse

Name:

Help feed a mouse by changing the size of a piece of cheese.

	Width of Hole	Width of Cheese	Comparison	Does the Mouse Eat?
a.				
b.				
c.				
d.				
e.				
f.				
g.				
h.				
i.				
j.				
k.				

Algebars: Exploring Properties of Operations

INTRODUCE

Project the sketch for viewing by the class. Expect to spend about 10 minutes.

1. Open **Algebars.gsp.** Go to page "Intro." Enlarge the document window so it fills most of the screen.

2. Explain, ***Today you're going to use Sketchpad to explore equivalent expressions. You'll use bars—called* algebars*—that represent variables and algebraic expressions.***

Review vocabulary as needed. A *variable* is a symbol, usually a letter, that represents a value that can change.

3. Press *Show Variables.* **The red bars are variables.** Drag point a and point b to the right and left to show how the lengths of the bars change. ***What is the value of a? What is the value of b?*** Have students practice reading the values of the variables as you drag the points to different locations. You might press *Show Vertical Lines* so that students can follow each gray vertical line up to find the value.

4. Drag point a to 0. ***What happens to the red bar?*** Students should observe that because the value is 0, the red bar disappears and becomes a point at 0.

An *algebraic expression* may contain variables, numbers, and operations.

5. Press *Show Algebars 1.* ***The green algebars are expressions. What expressions do these green bars represent?*** [$a + b$ and $b + a$] ***What do you think will happen to the lengths of the green bars when a is 1 and b is 1?*** Have students predict the lengths of each green bar. Then drag the points to check. Try other values for the variables, including negative values, and have students predict before you drag the points. Be sure students understand that the values of a and b are substituted into the expressions and affect their value. It's not necessary to point out that variables and numbers are also expressions.

6. ***What did you notice about the lengths of the two green bars? Are these two expressions equivalent?*** Drag the points again to change the values of a and b, if needed. Students should observe that for any values of a and b, the expressions are equivalent; the green bars are always the same length.

7. ***How could you indicate that these two expressions are equivalent?*** Students may make the following response: *Write it as an equation.* Write $a + b = b + a$ on the board after students suggest it.

8. Drag the points so that a and b are equal (a value close to 2 works well) and press *Show Algebars 2.* ***What do these green bars represent?*** [The expressions $a - b + 1$ and $b - a + 1$] ***What do you think will happen to the lengths of the green bars when I drag points a and b?***

Have volunteers share their thoughts, and then drag the points to several different values. Be sure to include other examples for $a = b$. ***What did you observe?*** Students should respond that the lengths of the green bars are the same only when $a = b$. ***Do you think that these expressions are equivalent? Can we use them to write an equation?*** Students should realize that because the green bars are not the same length all the time, the expressions are not always equal, so an equation cannot be written.

9. ***Now that you understand how algebars work, you can explore them on your own. You'll use them in this activity to explore some algebraic properties.***
10. If you want students to save their work in the Explore More, demonstrate choosing **File | Save As,** and let them know how to name and where to save their files.

DEVELOP

Expect students at computers to spend about 25 minutes.

11. Assign students to computers and tell them where to locate **Algebars.gsp.** Distribute the worksheet. Tell students to work through step 19 and do the Explore More if they have time.
12. Let pairs work at their own pace. As you circulate, here are some things to notice.
    - Encourage students to make predictions before dragging a and b. This will require them to think about how the variables and expressions are related.
    - In worksheet step 6, review the meaning of *commutative.* If an operation is commutative, it means that the operation can be performed in any order without affecting the outcome.
    - In worksheet step 7, as students are dragging, they should observe that the green bars for $a \div b$ and $b \div a$ will be equal lengths (that is, they both will be equal to 1) when $a = b$. ***Does this mean that the expressions are equivalent? Explain.*** This is a good check to be sure students understand that the green bars *always* need to be of equal length for the expressions to be equivalent.
    - In worksheet step 9, review the meaning of *associative.* If an operation is associative, changing the grouping of terms does not change the outcome.

- In worksheet steps 6 and 9, when the denominator is zero, the algebars disappear. Ask students why they think this happens with the model. Have them look at the expressions for division as a hint. It is not possible to divide by zero, so the algebars disappear whenever this happens.
- Review the order of operations as students work on the associative and distributive properties. ***What is the standard order of operations?*** [Operations within parentheses, exponents, multiplication and division from left to right, and then addition and subtraction from left to right] ***Why is it important for everyone to follow this order?*** [A standard rule ensures that everyone will get the same answer.]
- If students have time for the Explore More, they will learn how to construct algebars using **Custom** tools. After they build the algebars for the two expressions, students will explore whether they are equivalent. This is another investigation of the Distributive Property.

13. If students will save their work, remind them where to save it now.

SUMMARIZE

Project the sketch. Expect to spend about 10 minutes.

14. Gather the class. Students should have their worksheets with them. Open **Algebars.gsp** and go to page "Commutative." Begin the discussion by asking students to describe what equivalent expressions are. ***Today you used algebars to explore some equivalent expressions. What are equivalent expressions?*** Here are sample student responses.

    The green algebars, representing the expressions, stay the same length no matter where I drag the points representing the variables.

    They are expressions that are equal for any value of the variable.

15. Discuss worksheet steps 7, 10, 16, and 19 with the class. Have volunteers come up to the computer and prove which expressions are and are not equivalent. Summarize with the class by writing the properties on the board.

Property	Example
Commutative Property of Addition	a + b = b + a
Commutative Property of Multiplication	a • b = b • a
Associative Property of Addition	(a + b) + c = a + (b + c)
Associative Property of Multiplication	(a • b) • c = a • (b • c)
Distributive Property	a(b + c) = ab + ac

16. If time permits, discuss the Explore More. ***Were the expressions equivalent? If so, what equation did you write? Can we add this equation to our chart?*** Include $a(b - c) = ab - ac$ as another example of the Distributive Property. Explain that the Distributive Property works with subtraction as well because subtraction is the same as adding the opposite. Invite students to share other properties they may have explored.

EXTEND

1. Have students go to page "Make Your Own." Let students use the Sketchpad model to build algebars using the custom tools to discover whether these two expressions are equivalent: $2(a \div b)$ and $2a \div 2b$. Students will learn that the Distributive Property does not hold for multiplication over division. Have students share their thinking.
2. Have students explore which of the following expressions are equivalent: $a^c b^c$, $(a + b)^c$, and $(ab)^c$. They can build the expressions using the **(a^b)** tool. Students will learn that $a^c b^c = (a \cdot b)^c$ because exponents are distributive across multiplication, but not addition. Let volunteers share their results with the class.
3. ***What other questions might you ask about equivalent expressions or properties of operations?*** Encourage all inquiry. Here are some ideas students might suggest.

   If subtraction is the same as adding the opposite, why is subtraction not commutative?

   Is Corey's rule ever right? Are there any values of m and n for which $2 + mn = (2 + m)(2 + n)$*?*

   Why does multiplication distribute over addition?

   Why doesn't addition distribute over multiplication?

ANSWERS

3. The two expressions are equivalent because the green algebars always remain the same length. The equation is $a + b = b + a$.

5. The two expressions are not equivalent because the green algebars are not always the same length. They're the same only when $a = b$.

7. The addition algebars and the multiplication algebars are always the same length. The equations are $a + b = b + a$ and $ab = ba$.

8. Addition and multiplication are commutative; subtraction and division are not commutative.

10. The addition algebars and the multiplication algebars are always the same length. The equations are $(a + b) + c = a + (b + c)$ and $(a \cdot b) \cdot c = a \cdot (b \cdot c)$.

11. Addition and multiplication are associative; subtraction and division are not associative.

12. The expressions $2(c + 4)$ and $2c + 8$ are equivalent. As an equation, $2(c + 4) = 2c + 8$. You multiply 2 by each value in the parentheses and then add the results. This is an example of the distributive property of multiplication over addition.

13. Answers will vary. Some students may describe the behavior of the bars; others may give a counterexample; and others may give an algebraic argument in terms of the distributive property. Accept all reasonable answers. The main purpose is to get students to think about the question.

16. The two expressions are equivalent because the green algebars always stay the same length. The equation is $2(m + n) = 2m + 2n$. This is another example of the Distributive Property of Multiplication over Addition.

19. The two expressions, $2 + (m \cdot n)$ and $(2 + m) \cdot (2 + n)$, are not equivalent. The green algebars are not always the same length.

23. The two expressions are equivalent because the green algebars always stay the same length. The equation is $x(y - z) = xy - xz$. The Distributive Property works with subtraction as well because subtraction is the same as adding the opposite.

Algebars

Name:

In this activity you'll explore operations using algebars. Algebars are bars that represent algebraic quantities. Red bars are variables. Green bars are expressions. When two green bars are always equal in length, they represent equivalent expressions.

EXPLORE

1. Open **Algebars.gsp** and go to page "Intro." Press *Show Variables.* Two red algebars appear that represent the variables a and b. Drag points a and b left and right to change their values. You can press *Show Vertical Lines* to see their values more clearly.

2. Press *Show Algebars 1.* Two green algebars appear that represent the expressions $a + b$ and $b + a$. What do you think will happen to the lengths of the green algebars when you drag points a and b? Try it and check your prediction.

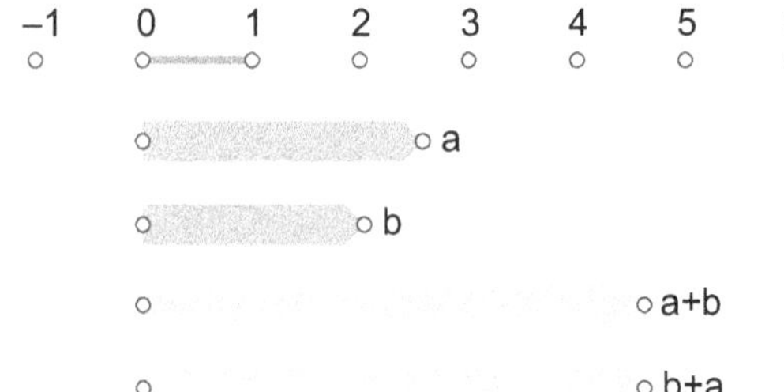

3. Are the two expressions equivalent? Explain. If so, write the result as an equation.

4. Press *Show Algebars 2.* Two green algebars appear that represent the expressions $a - b + 1$ and $b - a + 1$. What will happen to the green algebars when you drag points a and b? Try it and check your prediction.

5. Are the two expressions equivalent? Explain. If so, write the result as an equation.

Now you'll use algebars to explore some algebraic properties.

Commutative Property

6. Go to page "Commutative." This page shows four possible commutative properties.

Operation	Expressions
Addition	$a + b$ and $b + a$
Subtraction	$a - b$ and $b - a$
Multiplication	$a \cdot b$ and $b \cdot a$
Division	$a \div b$ and $b \div a$

Predict which green algebars will stay the same length when you drag points *a* and *b*. Then try it and observe what happens.

7. Which pairs of algebars are always the same length? Write an equation for each pair that matches.

8. Which of the four operations are commutative? Which are not?

Associative Property

9. Go to page "Associative." This page shows four possible associative properties.

Operation	Expressions
Addition	$(a + b) + c$ and $a + (b + c)$
Subtraction	$(a - b) - c$ and $a - (b - c)$
Multiplication	$(a \cdot b) \cdot c$ and $a \cdot (b \cdot c)$
Division	$(a \div b) \div c$ and $a \div (b \div c)$

Predict which green algebars will stay the same length when you drag points *a* and *b*. Then try it and observe what happens.

10. Which pairs of algebars are always the same length? Write an equation for each pair that matches.

11. Which of the four operations are associative? Which are not?

Distributive Property

12. Go to page "Distributive 1." Drag point c and observe what happens. Which two expressions are equivalent? Explain why. Write it as an equation.

13. Why is $2c + 4$ not equivalent to $2(c + 4)$?

14. Now go to page "Distributive 2." Sabrina says you can evaluate the expression $2(m + n)$ two different ways.

    $2(m + n)$: Add what's inside the parentheses first and then multiply by 2.

    $2m + 2n$: Multiply 2 by each value in the parentheses and then add the results.

15. Press *Show Sabrina's Algebars* to see how Sabrina tested her rule. The bottom two bars show the two methods.

16. Predict whether Sabrina is correct. Then drag points m and n and observe what happens. Are the two expressions equivalent? Explain. If so, write the result as an equation.

17. Corey says you can do something similar with the expression $2 + (m \cdot n)$.

    $2 + (m \cdot n)$: Multiply what's inside the parentheses first and then add 2.

    $(2 + m) \cdot (2 + n)$: Add 2 to each value in the parentheses and then multiply the results.

18. Press *Show Corey's Algebars* to see how Corey tested his rule. The bottom two bars show the two methods.

19. Predict whether Corey is correct. Then drag points m and n and observe what happens. Are the two expressions equivalent? Explain. If so, write the result as an equation.

EXPLORE MORE

Go to page "Explore More." Now you will build your own algebars to explore whether $x(y - z)$ and $xy - xz$ are equivalent expressions. You'll start by constructing the expression $(y - z)$.

20. Press and hold the **Custom** tool icon and choose the **(a−b)** tool. This tool requires you to click on five objects in this order:

- an unused white point
- the point at the tip of the y algebar
- the caption of this point (y)
- the point at the tip of the z algebar
- the caption of this point (z)

21. To complete the expression $x(y - z)$, choose the **(a∗b)** tool and click on five objects in order, as in step 20, using the x algebar and the $(y - z)$ algebar that you just constructed.

22. Next you'll construct the expression $xy - xz$. Start by constructing xy and xz, using the **(a∗b)** tool on the x, y, and z algebars. Then construct $xy - xz$ by using the **(a−b)** tool on the $(x \cdot y)$ and $(x \cdot z)$ algebars you just constructed.

23. Now drag points x, y, and z and observe what happens. Are the two expressions equivalent? Explain. If so, write the result as an equation.

24. Use the custom tools to create algebars of your own to test for other equivalent expressions, including ones with exponents.

2

Shapes, Graphs, and Data in Grade 6

Slanted Bases: Calculating Triangle Area 

ACTIVITY NOTES

INTRODUCE

Project the sketch for viewing by the class. Expect to spend about 10 minutes.

1. Introduce the activity. ***In this activity you'll explore the base, height, and area of a triangle that you'll construct.*** Review what students have learned about finding triangle area, including the formula $A = \frac{b \cdot h}{2}$. You might write the formula on the board. Don't bring up the idea that a triangle has three bases and three heights, but do entertain this point if students raise it, and let students know they'll be exploring this idea more using Sketchpad.

2. ***I'll model constructing a triangle in Sketchpad.*** Open page "Start" of **Slanted Bases Present.gsp** and enlarge the window so it fills most of the screen. As you demonstrate, make lines thick and labels large for visibility.

   - Use the **Segment** tool to draw the first side. Make the lines thick (**Display | Line Style**) for visibility.
   - Draw the other sides. Show students how to be sure that a vertex is highlighted when they click to connect another segment to it.
   - Drag a vertex to show that the triangle passes the drag test by remaining a triangle no matter how it is dragged.

3. ***You will be measuring the area of the triangle.***

   - If your students have not used the Calculator, model measuring the base of your triangle by selecting it and choosing **Measure | Length.** Then choose **Number | Calculate,** click on the measurement to insert it into the Calculator expression, and show students how to use the * on the Calculator to multiply.
   - In worksheet step 15, students check their calculations by measuring the area. Demonstrate how to construct the triangle interior by selecting the vertices, choosing **Construct | Triangle Interior,** and then choosing **Measure | Area.**

4. If students are to save their work, demonstrate how and where to save it.

DEVELOP

Expect students at computers to spend about 35 minutes.

5. Assign students to computers and distribute the worksheet. Tell students to work through all the steps of the worksheet and do the Explore More if they have time. Encourage students to ask a neighbor for help if they have questions about using Sketchpad.

If the labels appear automatically when you construct the segment, choose **Edit | Preferences.** In the Preferences dialog box, click the Text tab. For **Show labels automatically,** uncheck **For all new points.** Check **Apply to: New Sketches** and click **OK.** Then start again with a new sketch.

6. Let pairs work at their own pace. As you circulate, be alert to these issues.
   - In worksheet step 2, points may label automatically, depending on the setting in your software. Choosing **Edit | Preferences** will allow you to select to have Sketchpad show the labels of new points when they are created. Alternatively, students can use the **Text** tool to label points themselves.
   - In worksheet step 5, if you notice students are having trouble answering the question, ask, ***What do you notice about angles A and C when the perpendicular line falls outside the triangle?*** [One of these angles is obtuse.] Don't provide the answer. Students can move on if they don't see this right now.
   - In worksheet step 8, remind students to put their triangles to the drag test by dragging each of the vertices to make sure the line always intersects points *A* and *C.*
   - In worksheet step 11, if students have difficulty defining altitude, help them begin by providing the prompt, ***An altitude in a triangle is a line segment from . . . .*** Don't expect perfect definitions at this point. Students will have an opportunity to refine them in the whole-class discussion.
   - Worksheet step 17 illustrates an important objective of the activity, understanding that the base and altitude of a triangle need not be horizontal and vertical. Be sure students convince themselves of this fact by dragging their triangles sufficiently.

SUMMARIZE

Project the sketch for viewing by the class. Expect to spend about 15 minutes.

7. Students should have their worksheets. Open **Slanted Bases Present.gsp.** As you facilitate a class discussion of students' findings, invite volunteers to the computer to demonstrate.
8. Refer students to worksheet step 11. ***What are your ideas for a definition of altitude?*** [An altitude is a segment from the vertex of a triangle perpendicular to a line containing the base opposite that vertex.] Student definitions are apt to be less formal and may not include vocabulary such as *vertex* and *opposite.* However, make sure students understand the following two ideas.

- The altitude is perpendicular to the (line containing the) base.
- The altitude may not intersect the base, but instead may intersect the line extending the base.

A complete student definition will encompass answers to some of the previous worksheet questions.

9. ***What did you notice about the altitude's location and the type of angles (acute, obtuse, or right) at the triangle's base?*** Students should agree on the following points.
   - The altitude lies inside the triangle when both base angles are acute.
   - The altitude lies outside the triangle when a base angle is obtuse.
   - The altitude lies on a side of the triangle when a base angle is right.
10. ***Please share how you calculated the area of the triangle (step 13), and explain why that calculation worked.*** Students should have drawn on their previous understanding of the triangle area formula, $A = \frac{base \cdot height}{2}$, or $A = \frac{1}{2}base \cdot altitude$. They should have entered into the Calculator $\frac{1}{2}m\overline{AC} \cdot m\overline{BD}$ or $\frac{m\overline{AC} \cdot m\overline{BD}}{2}$.

You may also ask whether a triangle needs a "bottom" segment.

11. ***Does the base of a triangle have to be the bottom segment? What did you do today to answer this question?*** The students' discussion should include two observations.
   - Students moved the triangle's base and altitude to different positions.
   - Students constructed three altitudes, understanding that any side could serve as a base of the triangle.
12. Invite students who have completed the Explore More to share their additional observations about the three altitudes in a triangle.

ANSWERS

1. Most students will show the vertical altitude.
5. The perpendicular line doesn't intersect base $\overline{AC}$ when either $\angle A$ or $\angle C$ is obtuse or measures more than 90°.
6. The perpendicular line passes directly through point A when $\angle A$ is a right angle and directly through point C when $\angle C$ is a right angle.

11. An *altitude* in a triangle is a segment from the vertex of the triangle to a line containing the base opposite that vertex; the segment is perpendicular to that line.

15. Answers will very.

16. $A = \frac{base \cdot altitude}{2}$, or $A = \frac{1}{2} base \cdot altitude.$

17. The base is vertical.

22. The three calculations are equal. The area of a triangle can be calculated using any side as its base.

23. Students should identify all three sides of the triangle as bases and draw a corresponding altitude for each base.

24. Answers will vary depending on students' understanding at the beginning of the activity. Students should have learned that any side of a triangle can be a base, that for each base there is a corresponding altitude, and that an altitude can be outside the triangle.

25. In an acute triangle, the altitudes all lie inside the triangle. In an obtuse triangle, two altitudes lie outside the triangle. In a right triangle, two altitudes lie on the two sides forming the right angle, and the third lies inside. It is not possible for exactly two altitudes to lie inside the triangle; when one altitude is outside, another one is as well.

26. The lines through the three altitudes always intersect in a single point. This point, called the *orthocenter,* lies outside an obtuse triangle, inside an acute triangle, and at the vertex of the right angle in a right triangle.

Slanted Bases

Name:

In this activity you will explore the bases, heights, and areas of triangles.

1. Using a pencil, draw any heights and mark any bases on this triangle.

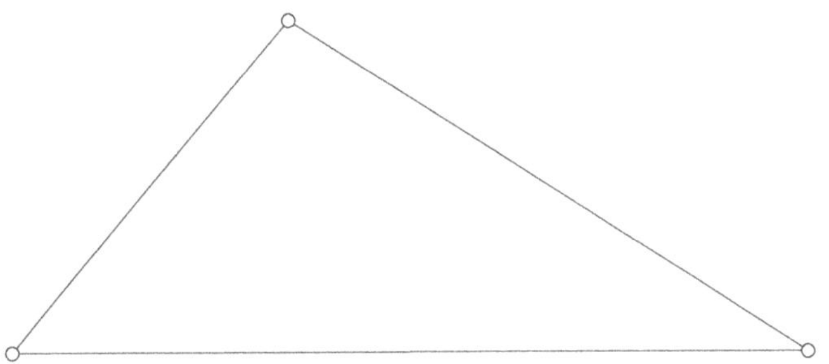

CONSTRUCT

2. Construct a triangle.

Label the vertices *A*, *B*, and *C*.

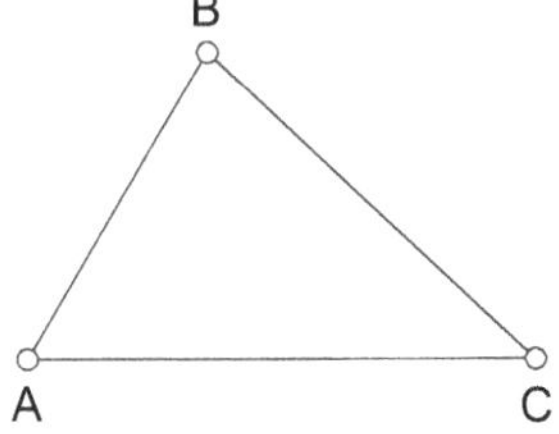

3. To construct an altitude from point *B* to base $\overline{AC}$, select *B* and $\overline{AC}$ and choose **Construct | Perpendicular Line.**

   The height of the triangle from point *B* to base $\overline{AC}$ is measured along this perpendicular line.

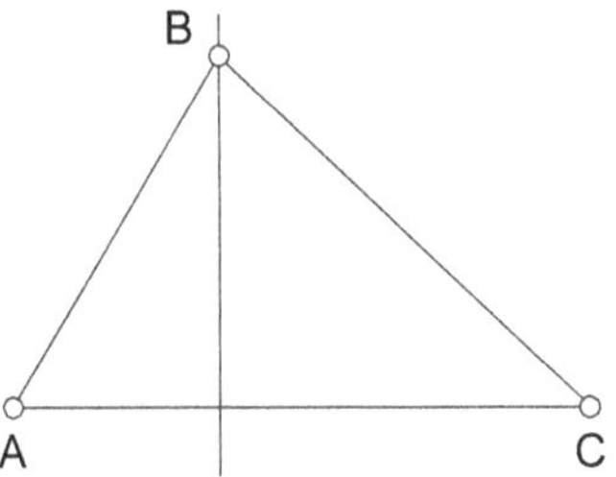

4. Drag point *B*. Note that the perpendicular line sometimes intersects base $\overline{AC}$ and sometimes doesn't.

5. What do you notice about the triangle when the perpendicular line doesn't intersect the base?

6. Describe the triangle when the perpendicular line passes directly through point *A* or point *C*.

Slanted Bases

continued

7. Drag point *B* so that the line does not intersect the base.

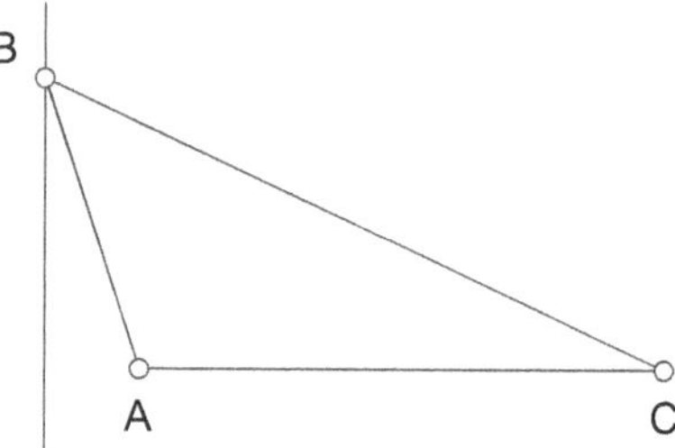

8. The perpendicular line and the base no longer intersect. To make them intersect again, you need to extend the base.

   Construct a line through points *A* and *C*.

   Drag the vertices to make sure the base and the perpendicular line always intersect.

9. Select the perpendicular line and the extended base $\overleftrightarrow{AC}$ and choose **Construct | Intersection.**

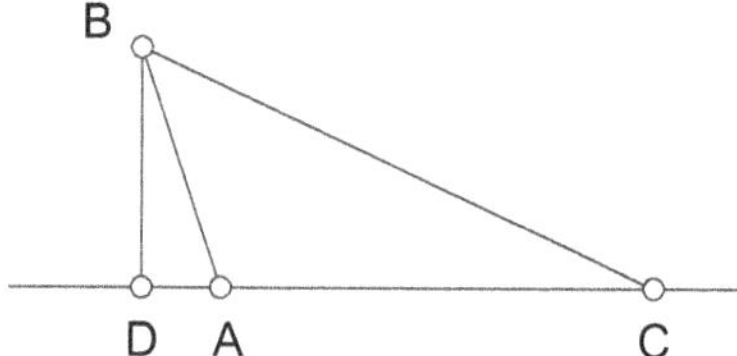

   Label the point of intersection *D*.

10. Hide the perpendicular line by selecting it and choosing **Display | Hide Perpendicular Line.**

   Construct a line segment between points *B* and *D*.

11. Segment *BD* is an *altitude* of the triangle.

   Use what you know about constructing $\overline{BD}$ to define *altitude*.

12. Measure the length of altitude $\overline{BD}$ by selecting it and choosing **Measure | Length.** Also measure the length of base $\overline{AC}$.

13. Now you will use the triangle area formula to find the area of this triangle. Choose **Number | Calculate.**

   Click on measurements in your sketch to insert them into the Calculator. (You may need to drag the Calculator to the side.) Then click **OK.**

14. Hide the measurements (not the calculation) by selecting them and choosing **Display | Hide Measurements.**

15. Now you will check your calculation.

    Select points *A*, *B*, and *C* and choose **Construct | Triangle Interior.**

    With the interior still selected, choose **Measure | Area.**

    Does the area measurement match your calculation? ____________

    If not, double-click your calculation. Change it so that it is correct. Drag vertices to confirm the two area measurements are the same.

16. Write the formula that tells how you calculated the area. Use *base* for the length of base $\overline{AC}$ and *altitude* for the length of $\overline{BD}$.

    Area = ____________

17. Drag the vertices and edges of the triangle.

    Sketch a triangle with its altitude horizontal.

    When the altitude is horizontal, what can you say about the base?

    Sketch a triangle that has its base on the top with the triangle pointing down.

Look at the questions in step 17 another way.

18. Use $\overline{AB}$ as the base and construct an altitude to it from vertex *C*. Start by extending the base, and use the same process you did before.

19. Measure altitude $\overline{CE}$. Measure base $\overline{AB}$.

    Use these lengths to calculate the area of the triangle.

    Hide the altitude and base measurements, but don't hide the calculations.

20. Now use $\overline{CB}$ as the base and construct an altitude $\overline{AF}$ to the extension of base $\overline{CB}$.

21. Measure the lengths of altitude $\overline{AF}$ and base $\overline{BC}$.
    Use these lengths to calculate the area of the triangle.

22. What do you notice about the area calculations? Explain what this shows.

23. Using a pencil, draw any bases and heights in the triangle below.

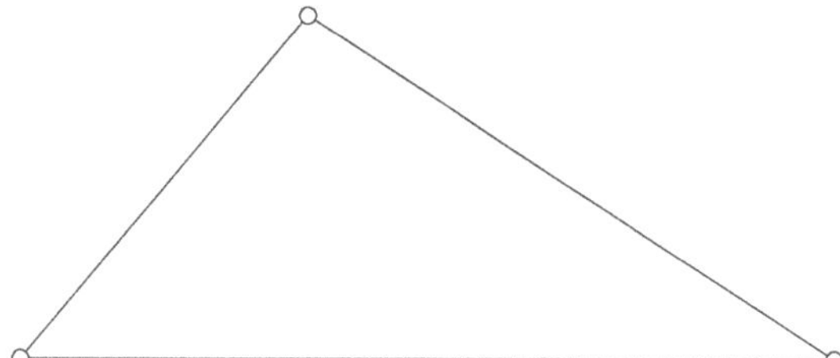

24. Compare your drawing in step 23 to your drawing in step 1. What have you learned?

EXPLORE MORE

25. As you make the Sketchpad triangle acute, obtuse, and right, watch the three altitudes. Write about your findings. Can you make a triangle in which exactly two of the altitudes fall inside the triangle? Explain.

26. Make your triangle obtuse and extend the three altitudes by constructing lines. What do you notice? The point where the altitudes intersect is called the *orthocenter* of the triangle. Write some observations about where the orthocenter is located. Consider obtuse, acute, and right triangles.

One Parallel Pair: Trapezoid Area

INTRODUCE

Project the sketch for viewing by the class. Expect to spend about 10 minutes.

1. Open Sketchpad and enlarge the document window so it fills most of the screen.
2. Explain, ***Today you're going to use Sketchpad to discover the area of a trapezoid. First let's define* trapezoid. *What is a trapezoid?*** Listen for the following responses.

   A trapezoid is a plane figure; it's a polygon.

   A trapezoid has four sides; it's a quadrilateral.

   A trapezoid has two sides that are parallel.

   The parallel sides are called bases.

   The two nonparallel sides are called legs.

   Is a parallelogram a trapezoid? Explain. [Yes, if you use an inclusive definition of trapezoid (having at *least* one pair of parallel sides); no, if you use an exclusive definition of trapezoid (having *exactly* one pair of parallel sides).] Write a definition for *trapezoid* on chart paper. Here is a sample definition: A trapezoid is a quadrilateral that has one pair of parallel sides. The parallel sides are called *bases* and the nonparallel sides are called *legs.*
3. ***I'll demonstrate how to construct a trapezoid using the definition we wrote.*** As you demonstrate, make lines thick and labels large for visibility. Model the trapezoid construction in worksheet steps 1–8. Here are some tips.
   - Start by choosing **Edit | Preferences | Text** and choosing **Show labels automatically: For all new points.** Explain that for purposes of the demonstration, points will be automatically labeled as you construct the trapezoid. Tell students that they will label their points using the **Text** tool after they construct the trapezoid.
   - In worksheet steps 1 and 2, talk about the parts of the trapezoid as you construct them. ***Segment AB will be one of the bases of the trapezoid. Segment AC will be one of the legs. What will I need to do next to construct the other base?*** [Make a line parallel to $\overline{AB}$ through point *C*.] Construct the parallel line.
   - In worksheet step 4, ask students whether it matters where point *D* is placed. ***According to our definition, does it matter where I construct***

point D on the line? [Yes, point *D* needs to be to the right of point *C* for the shape to have four sides.]

- Follow worksheet steps 6–8 to finish constructing the trapezoid. ***How does this shape meet our definition of a trapezoid?*** [The shape has four sides and one pair of parallel sides, $\overline{AB}$ and $\overline{CD}$.] ***What is the name of this shape?*** [Trapezoid *ABDC*] Remind students to name a trapezoid by listing the vertices in consecutive order.
- Model how to choose **Edit | Undo.** Explain that students can use this command to "undo" a previous action if they make a mistake in their construction.

4. If you want students to save their work, demonstrate choosing **File | Save As,** and let them know how to name and where to save their files.

DEVELOP

Expect students at computers to spend about 20 minutes.

5. Assign students to computers. Distribute the worksheet. Tell students to work through step 20 and do the Explore More if they have time. Encourage students to ask their neighbors for help if they are having difficulty with the construction.
6. Let pairs work at their own pace. As you circulate, here are some things to notice.
   - In worksheet step 11, check students' understanding of the term *height.* ***Can you explain what the height of the trapezoid is?*** [It is the perpendicular distance from one base to the other base.] Explain that when Sketchpad measures a distance between a point and a line, it measures the shortest distance, or the length of a perpendicular line segment from the point to the line. For your understanding, height is also referred to as *altitude.*
   - In worksheet step 12, listen to students as they drag different parts of the trapezoid. Ask questions to check their understanding. ***What happens to your measurements when you drag point D to the right?*** [The base $\overline{CD}$ gets longer, the base $\overline{AB}$ and the height stay the same, and the area gets larger.] ***What happens when you drag $\overline{CD}$ up?*** [The lengths of the bases stay the same, while the height and area increase.] At this point, students only observe that changing the length of the base changes the area and changing the height changes the area.

- In worksheet step 14, tell students that point E will flash briefly to indicate that it has been marked as a center. ***This point will be the center of the rotation.*** Have students try to guess what shape the trapezoid and its rotated image will form before they rotate the trapezoid. For your understanding, students will need to select points A and C as well as the interior of the trapezoid, so that points A' and C' appear in the rotated image. Remind students that in the rotated shape, the corresponding point to A is called "A prime" and is written A'. The corresponding point to C is C'.
- In worksheet step 16, students should recognize the new shape as a parallelogram. Have students identify the parts of the parallelogram. ***What is the base of the new shape?*** [The length of $\overline{AC'}$ or length of $\overline{CA'}$] ***Is $\overline{DB}$ the height of the new shape?*** [No] ***How do you know?*** [It is not perpendicular to the base.] Note that the height needs to be constructed.
- In worksheet step 17, help students as needed. ***What is the length of $\overline{BC'}$ equal to?*** [It is the same length as $\overline{CD}$.]
- In worksheet step 18, students will need to recall the formula for the area of a parallelogram. ***What is the area formula for the new shape?*** [$A = bh$]
- Check that students use parentheses around the sum of the bases. If not, try to get them to recognize their error. ***Is $4 + 6 \div 2$ the same as $(4 + 6) \div 2$? Explain.*** [No; following the order of operations, the value of the first expression is 7 and the value of the second expression is 5.]
- If students have time for the Explore More, they will derive another formula for the area of a trapezoid using the median.

7. If students will save their work, remind them where to save it now.

SUMMARIZE

Project the sketch. Expect to spend about 15 minutes.

8. Gather the class. Students should have their worksheets with them. Open **One Parallel Pair Present.gsp** and go to page "Trapezoid." Begin the discussion by having students identify the new shape formed by the transformation. ***What new shape was formed by the trapezoid and its rotated image?*** [Parallelogram $AC'A'C$] ***How do you know?*** [The shape is a parallelogram because opposite sides are parallel.]

9. Have students explain how they arrived at their answers to worksheet step 17. ***What is the base of the new shape if b_1 represents the length of $\overline{AB}$ and b_2 represents the length of $\overline{CD}$? Explain your reasoning.*** Students should reason that $\overline{CD}$ and $\overline{BC'}$ are the same length; $\overline{BC'}$ is just the rotated image of $\overline{CD}$. The base of the parallelogram is $m\overline{AB} + m\overline{BC'}$ or $m\overline{AB} + m\overline{CD}$ or $b_1 + b_2$.

10. For worksheet step 18, ask volunteers to share their answers. ***What area formula did you use?*** [Area of a parallelogram, $A = (b_1 + b_2)h$] ***Why are the parentheses important in your formula?*** [You need to add the two base measurements before multiplying by the height.]

11. For worksheet step 19, ask students for the formula of a trapezoid. ***How did you arrive at your formula?*** Students' reasoning may vary. Here are some possible replies.

    It took two trapezoids to make the parallelogram, so the area of one trapezoid is one-half the area of the parallelogram. So it is $bh \div 2$, where b is the sum of the two bases of the trapezoid, or $(b_1 + b_2)h \div 2$.

    The heights of the trapezoid and the parallelogram are the same. The base of the parallelogram is equal to both bases of the trapezoid added together. So we substituted $b_1 + b_2$ for b in the area formula for a parallelogram and got $h(b_1 + b_2)$. That's the area of the parallelogram. But two trapezoids formed the parallelogram, so to find the area of one trapezoid, we divided by 2 and got $h(b_1 + b_2) \div 2$.

12. Review worksheet step 20 by pressing *Show Calculation* on the sketch. Have students verify that their expressions match. Point out that the value of the expression and the area of trapezoid *ABDC* are equal.

13. Go to page "Double Trapezoid" and press *Double Trapezoid* to show another example of the relationship between the area of a trapezoid and the area of a parallelogram formed by the trapezoid and its rotated image. It is clear in this sketch that the new shape has a base measurement of $b_1 + b_2$, which are the two bases of the trapezoid.

14. If time permits, discuss the Explore More. ***How is the midsegment related to the two bases?*** Students should understand that the midsegment (median) of a trapezoid is also the average length of the two bases, so the height times the midsegment (median) length is equal to the area.

ANSWERS

16. The two combined trapezoids form a parallelogram.
17. The parallelogram formed by the combined trapezoids has a base of length $b_1 + b_2$.
18. $A = (b_1 + b_2)h$ is the area of the parallelogram.
19. $A = \frac{1}{2}(b_1 + b_2)h$ is the area of the trapezoid.
20. $\dfrac{(m\overline{AB} + m\overline{CD})(Distance\ C\ to\ \overline{AB})}{2}$

    Students must use parentheses when they sum the two bases.
21. $A = mh$, where m is the length of the midsegment, gives the area of the trapezoid. The length of the midsegment is the average of the lengths of the bases.

One Parallel Pair

Name:

In this activity you'll construct a trapezoid and then transform it into a familiar shape. You'll use the formula for the area of this shape to write the formula for the area of a trapezoid.

CONSTRUCT

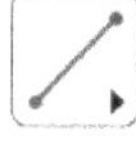

1. Construct two segments that share an endpoint.

2. Label the shared endpoint *A* and the other two endpoints *B* and *C*.

3. Now you'll construct a line through point *C* parallel to $\overline{AB}$.
   Select point *C* and $\overline{AB}$ and choose **Construct | Parallel Line.**

4. Construct point *D* on the line and label it.

5. Construct $\overline{BD}$.

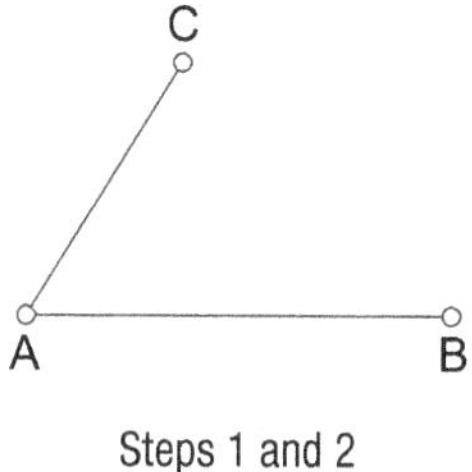

Steps 1 and 2

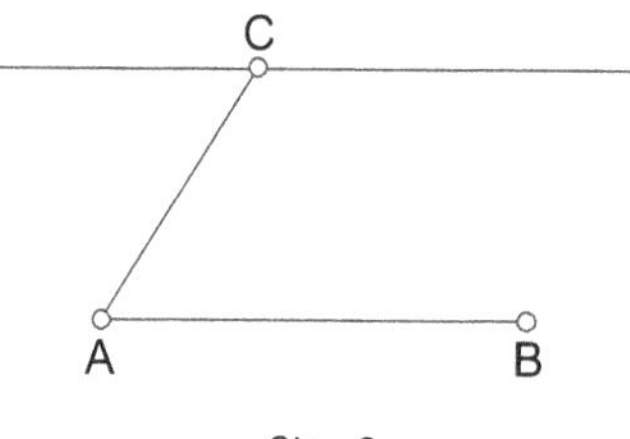

Step 3

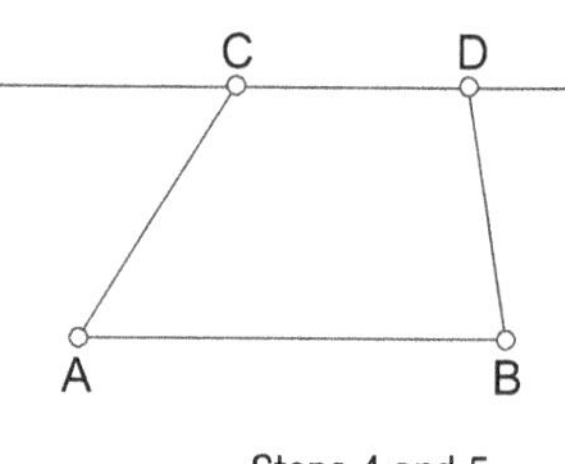

Steps 4 and 5

6. Hide the line by selecting the line and choosing **Display | Hide Parallel Line.**

7. Construct $\overline{CD}$.

8. Next, you'll construct the interior of trapezoid *ABDC*.
   Select the vertices in consecutive order and then choose **Construct | Quadrilateral Interior.**

EXPLORE

Area ABDC = 4.20 cm^2
m $\overline{AB}$ = 3.78 cm
m $\overline{CD}$ = 2.15 cm
Distance C to $\overline{AB}$ = 1.42 cm

C D
A B

9. Measure the area of *ABDC* by selecting the interior and choosing **Measure | Area.**

10. You will measure the lengths of the bases of the trapezoid, $\overline{AB}$ and $\overline{CD}$.
    Select each base and choose **Measure | Length.**

11. Now you'll measure the height of the trapezoid.
    Select point *C* and $\overline{AB}$. Then choose **Measure | Distance.**

12. Drag different parts of the trapezoid and observe the measures.

At this point, it's probably hard to see any relationships between the area measure and the base and height measurements.

CONSTRUCT

13. Select $\overline{DB}$ and choose **Construct | Midpoint.**
    Label the midpoint *E*.

14. Now you'll mark point *E* as a center of rotation and rotate the entire trapezoid by 180°.
    Double-click point *E* to mark it as a center.
    Select points *A* and *C* and then the interior of the trapezoid.
    Choose **Transform | Rotate.**
    In the dialog box, enter 180 for Angle.
    Click **Rotate.**

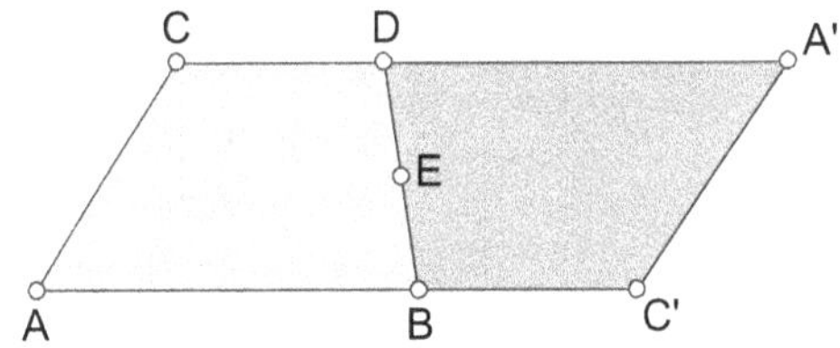

EXPLORE

15. Drag parts of the figure and observe the shape formed by the trapezoid and its rotated image.

16. What shape do the two combined trapezoids form?

17. Let b_1 represent the length of base $\overline{AB}$ and let b_2 represent the length of base $\overline{CD}$. What is the length of the base of the shape formed by the combined trapezoids?

18. Write a formula for the area of the combined shape in terms of b_1, b_2, and h (for height).

One Parallel Pair

continued

19. Write a formula for the area of a single trapezoid in terms of b_1, b_2, and h.

20. In your sketch, check that you've derived the correct formula by calculating an expression equal to the area of the trapezoid. Use $m\overline{AB}$, $m\overline{CD}$, and the distance from point C to $\overline{AB}$ in your expression.

    Choose **Number | Calculate** to open the Sketchpad Calculator.

    Click once on a measurement to enter it into a calculation. Use parentheses where necessary. Record your expression.

EXPLORE MORE

21. Construct the midpoints of the nonparallel sides of the trapezoid. Connect these midpoints with a segment. Use the length of this midsegment to invent a new area formula.

Smoothing the Sides: Regular Polygon and Circle Area

ACTIVITY NOTES

INTRODUCE

Project the sketch for viewing by the class. Expect to spend about 10 minutes.

1. Open **Smoothing the Sides.gsp** and go to page "Polygon." Enlarge the document window so it fills most of the screen.
2. Explain, ***Today you're going to use Sketchpad to explore the areas of regular polygons. Based on your investigations, you'll write area formulas for any regular polygon, and then for a circle. What is a regular polygon?*** Work with students to come up with a common definition. Here is a sample definition: A regular polygon is a polygon that has all congruent sides and all congruent angles.
3. ***What is the name of this regular polygon?*** [Regular hexagon] Remind students, if needed, that polygons are named by the number of sides. Review the names of common regular polygons.

Number of Sides	Regular Polygon
3	equilateral triangle
4	square
5	regular pentagon
6	regular hexagon
7	regular heptagon
8	regular octagon
9	regular nonagon
10	regular decagon
12	regular dodecagon

4. ***Suppose I draw a segment from the center of a regular polygon to each vertex. What shapes would I form?*** [Triangles] Use the **Arrow** tool to select the segments with a thick outline on the sketch. ***Into how many triangles is this regular hexagon divided?*** [Six] ***Are these congruent triangles? How do you know?*** [Yes, each has sides of the same length.] ***Into how many congruent triangles do you think a square can be divided?*** [Four] Let students guess, and then drag the marker to 4 to confirm. Explain that by dragging the marker you can change the number of sides of the regular polygon. Let students explore this relationship further on their own.
5. Review what *s* and *a* represent in the sketch. ***What does s represent?*** [Side length] ***What does a represent?*** [Apothem] Students may say that

this is the height of the triangle, which is true. Explain that it is also an *apothem* of the regular polygon, and provide a definition. Here is a sample definition: An apothem of a regular polygon is the line segment from the center to the midpoint of a side.

6. Tell students that they will now explore the sketch on their own to write area formulas for any regular polygon and for a circle.

DEVELOP

Expect students at computers to spend about 25 minutes.

7. Assign students to computers and tell them where to locate **Smoothing the Sides.gsp.** Distribute the worksheet. Tell students to work through step 14 and do the Explore More if they have time. Encourage students to ask their neighbors for help if they are having difficulty with the exploration.

8. Let pairs work at their own pace. As you circulate, here are some things to notice.

   - In worksheet step 4, students should already know that the formula for area of a triangle is $\frac{1}{2}bh$. Be alert to students who don't; they will need a quick review.
   - In worksheet steps 4 through 8, have students use the Sketchpad Calculator to verify their formulas. ***How can you test that your formula is correct?***
   - For worksheet steps 10 and 11, ask students who need help to observe the relationships among the measurements as the number of sides increases. ***Are any measurements almost the same?***
   - In worksheet step 13, have students who are having trouble review their answer for worksheet step 8. ***What formula did you write for the area of the regular polygon, using its apothem and perimeter? How do these measurements compare to the radius and the circumference of the circle?***
   - If students have time for the Explore More, they will investigate whether the ratio of the regular polygon's perimeter to twice its apothem approaches a familiar number, *pi.* Then students will write this ratio using the circumference and the diameter of the circle.

SUMMARIZE

Project the sketch. Expect to spend about 10 minutes.

9. Gather the class. Students should have their worksheets with them. Begin the discussion by reviewing worksheet step 4. ***What does "a" represent in the triangle?*** [Height] ***What does "s" represent in the triangle?*** [Base] ***What is the formula for the area of the triangle?*** $\left[\frac{1}{2}bh\right]$ ***What is the formula using "a" and "s"?*** $\left[\frac{1}{2}sa\right]$

10. Review worksheet steps 5 and 6. ***How many triangles are formed?*** [4 in the square; 5 in the pentagon] ***Are they all congruent?*** [Yes] ***How can you use this to find the area of each regular polygon?*** [Multiply the area of one triangle by the number of sides] Make sure students see that joining every vertex of any regular polygon to the center point forms as many isosceles triangles as there are sides of the polygon.

11. Now review worksheet step 7. ***What is the formula for the area of a regular polygon with n sides? Explain your reasoning.*** $[A = \frac{1}{2}asn]$ Students may come up with the following responses.

    The number of sides of a regular polygon is the same as the number of congruent triangles that make up the regular polygon. So you multiply the area of one triangle by the number of sides to find the total area.

    We found the area by first finding the total length of all the bases of the triangles. We did this by multiplying the base (or side length) by the number of sides, n. Then we multiplied by the height, or apothem, and divided by two.

    We discovered that you can construct n congruent triangles in a regular polygon with n sides. So, the formula for a regular polygon is the area of one triangle multiplied by n, the number of sides.

12. Review worksheet step 8, asking students to explain their formulas. ***How is the perimeter related to the variable s?*** $[P = sn]$ Make sure students understand that the variable P is being substituted into the area formula. This is necessary to derive the circle formula.

13. For worksheet steps 9–12, drag the marker to increase the number of sides of the regular polygon. ***What does the regular polygon look like now?*** [When the number of sides increases to 18 or greater, the regular polygon resembles a circle.] ***What does the apothem approach?*** [Radius] ***How about the perimeter?*** [Circumference] As the measurements get closer to each other, have students read them aloud.

14. For worksheet step 13, ask students to explain their answers. Students may reply with the following explanations.

    In step 8, we wrote the area formula for the regular polygon as one-half the apothem multiplied by the perimeter. As the regular polygon approaches a circle, the apothem approaches the radius and the perimeter approaches the circumference. We substituted the radius for the apothem and the circumference for the perimeter and came up with one-half the radius multiplied by the circumference.

    As the number of sides increases, the area of the regular polygon gets closer and closer to the area of the circle, the perimeter of the regular polygon gets closer and closer to the circumference of the circle, and the apothem gets closer and closer to the radius. We figured out that the area of the regular polygon is one-half its apothem multiplied by its perimeter, so the area of the circle is one-half the radius multiplied by its circumference.

15. In worksheet step 14, have students give the area for a circle after substituting and simplifying.

16. If time permits, discuss the Explore More. ***If the apothem approaches the radius as the number of sides increases, what does twice the apothem approach?*** [Diameter] ***What does the ratio of the perimeter to twice the apothem approach?*** [π] ***How can you write this ratio using the circumference and the diameter of the circle?*** [C/d]

ANSWERS

4. $A = \frac{as}{2}$ or $A = \frac{1}{2}as$
5. $A = 4\left(\frac{as}{2}\right)$ or $A = 4\left(\frac{1}{2}as\right)$
6. $A = 5\left(\frac{as}{2}\right)$ or $A = 5\left(\frac{1}{2}as\right)$
7. $A = n\left(\frac{as}{2}\right)$ or $A = \frac{1}{2}asn$
8. The perimeter is equal to s times the number of sides: $P = sn$

   $A = \frac{1}{2}aP$
9. As the number of sides increases, the polygon becomes more and more like a circle.
10. The apothem approaches the radius of the circle.
11. The perimeter approaches the circumference of the circle.

12. The area of the polygon approaches the area of the circle.
13. $A = \frac{1}{2}rC$
14. $A = \frac{1}{2}r(2\pi r)$

    $A = \pi r^2$
15. The ratio approaches the measurement for *pi*, 3.14 . . . .
16. $\frac{C}{d}$

Smoothing the Sides

Name:

In this activity you'll figure out how to calculate the area of any regular polygon by dividing it into congruent isosceles triangles. Then you'll explore how you can use this approach to calculate the area of a circle.

POLYGON AREA

1. Open **Smoothing the Sides.gsp** and go to page "Polygon."

s = 5.64 cm

a = 4.88 cm

Area hexagon = 82.651 cm^2

a

s

sides

2 3 4 6 8 12 18 24

2. Drag the *sides* marker along its segment. Observe how the polygon and the measurements change.

3. Each regular polygon with four or more sides can be divided into triangles. Look at the triangle with a thick outline. It has segments labeled *a* for *apothem* and *s* for *side.* The side is the base of this triangle, and the apothem is the height of the triangle.

4. Go to page "Triangle," where you'll find an isosceles triangle. Choose **Number | Calculate** and write an expression to find the area of the triangle using *s* and *a.* Click on the values in the sketch to enter them into the Calculator. Check your expression against the measured area. Drag the vertices of the triangle to make sure your expression remains equal to the measured area for any isosceles triangle. What is your expression?

Smoothing the Sides

continued

5. Go to page "Square." Press *Show Triangles* and then repeat step 4 to find the area of any square. What is your expression?

6. Go to page "Pentagon" and repeat step 5 to find the area of any regular pentagon. What is your expression?

7. Write an expression for the area of any regular polygon. Use n to represent the number of sides, as well as the variables s for the side length and a for the apothem length.

8. How is the perimeter of the polygon related to the variable s? Write another expression for the area of any regular polygon with perimeter P and apothem length a.

CIRCLE AREA

s = 5.64 cm
a = 4.88 cm
Area hexagon = 82.651 cm^2
Area circle = 99.94 cm^2
r = 5.64 cm
Circumference = 35.44 cm

sides
2 3 4 6 8 12 18 24

9. Go to page "Circle." Drag the *sides* marker. What happens to the polygon as the number of sides increases?

Smoothing the Sides

continued

10. What measure of the circle does the apothem approach as the number of sides increases?

11. What measure of the circle does the perimeter approach as the number of sides increases?

12. How does the measured area of the polygon compare to the measured area of the circle as the number of sides increases?

13. Write a formula for the area of a circle, similar to the formula you wrote in step 8, using C for circumference and r for radius.

14. The formula for circumference is $C = 2\pi r$. Substitute $2\pi r$ for C in your formula in step 13 and simplify.

EXPLORE MORE

15. Use the sketch to explore the ratio $\frac{p}{2a}$ for the regular polygon as the number of sides increases. What measurement does this ratio approach?

16. Write this ratio using C for circumference and d for diameter of the circle.

INTRODUCE

Project the sketch for viewing by the class. Expect to spend about 15 minutes.

1. If centimeter cubes are available, distribute about 25 (at least 20) to each student. Open **Stack It Up.gsp.** Go to page "Layers." ***You see one cube.*** Drag point *L* to add one cube at a time until there are five cubes.

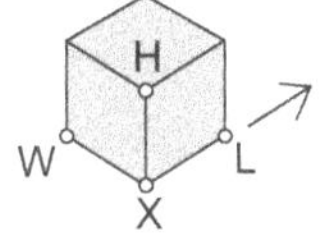

   What is happening? Introduce the language, *row* of cubes, if students don't. For students who need help visualizing the two-dimensional representation as a three-dimensional shape, building this row with centimeter cubes will be helpful.

2. Drag point *W* so that the row of cubes grows to a layer that has 2 rows of 5 cubes, then 3 rows of 5 cubes, and finally 4 rows of 5 cubes. If students have centimeter cubes, have them represent the growing layer using the cubes. With the addition of each row, ask students what is happening. By focusing on the growth pattern, you'll help students to interpret what they see and to develop a deep understanding of volume.

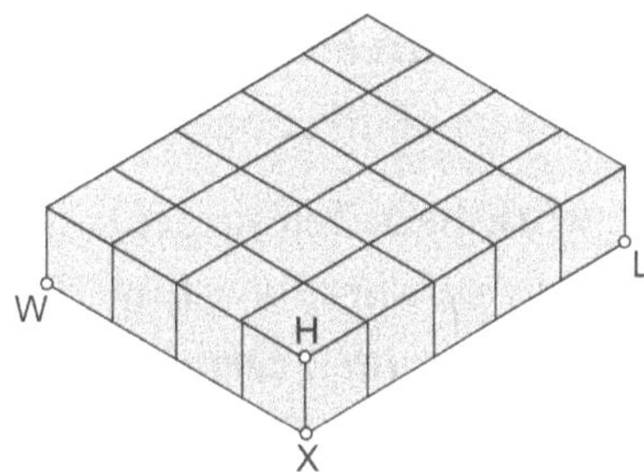

3. ***Let's call this one layer. How many cubes fill this layer? How do you know?*** Elicit the idea that there are 4 rows of 5 cubes, or $4 \times 5 = 20$ cubes. Because there is only one layer, some students may confuse the answer, 20 cubes, with the measure of the *area* of the top face of the layer. Clarify that 20 cubes tells about the amount of space filled, while area is a measure of the amount of space covered. The area of the top face of the layer is 20 square centimeters. Cut a 4-by-5 rectangle from centimeter graph paper and ask whether this is equal to the space filled by the cubes in the model. [It isn't.]

4. Drag point *H* so that the single layer of cubes grows to 2 layers of 20 cubes each, and then to 3 layers of 20 cubes each. Ask students what is happening.

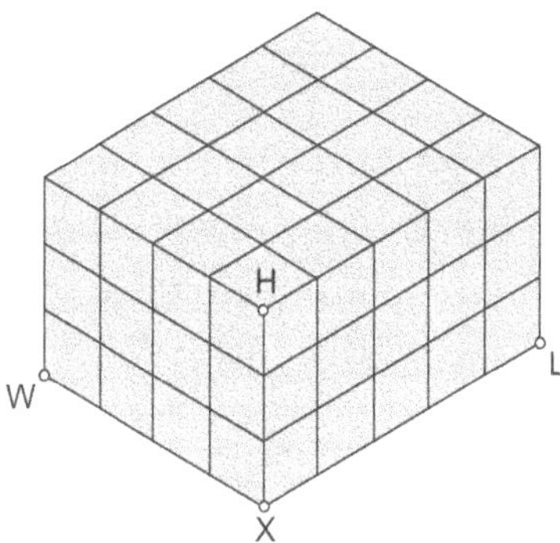

More layers are showing up. They are stacked on top of each other.

The first layer is being copied.

Dragging point H increases the height of the box.

Do not introduce the formula for volume of rectangular prisms (Volume = length × width × height).

How many cubes fill this box? How do you know? Students may explain that there are 3 layers of 20 cubes, or 3 × 20 = 60 cubes.

5. Distribute the worksheet. Tell students to work through steps 1–6 and do the Explore More if they have time. ***In each step, you will use the model to make boxes with a certain number of cubes.*** Model enlarging the window so it fills most of the screen. Drag point *X* and let students know they can move a box to make room to build large boxes. ***The largest box that can be built using the model is 12 cm long, 12 cm wide, and 12 cm tall.***

6. If students will save their work, model choosing **File | Save As** and tell how they should name their files and where to save them.

DEVELOP

Expect students at computers to spend about 30 minutes.

7. Assign students to computers and tell them where to locate **Stack It Up.gsp.** Encourage students to ask a neighbor for help using Sketchpad if needed.

8. Let pairs work at their own pace. As you circulate, observe and listen to students' conversations. Here are some things to notice.

    - In worksheet step 1, students will probably take note of the values for *XL, XW,* and *XH* that appear and update automatically in the sketch. Let students explore these on their own. Some students may begin to make sense of the displayed values and use them; other students may disregard them.
    - In worksheet step 1, some students may not realize that a box with a single row of cubes is one possible solution; they may think that all three dimensions need to be larger than 1 cm.
    - In worksheet steps 1 and 2, challenge students to make a box with no side of length 1 cm. Also notice any students who make the conjecture that the lengths of *XL, XW,* and *XH* must all be factors of the total number of cubes.
    - In worksheet steps 5 and 6, students are asked to build two different boxes using the same number of cubes in each box. It's fine if students make boxes that share one dimension, such as a 2 cm × 4 cm × 3 cm box and 2 cm × 1 cm × 12 cm box. Prompt those students to extend their thinking by saying, ***I wonder if it's possible to make two boxes so that they have no lengths in common.***
    - In the Explore More steps, students are given several new problem types. Encourage students to persevere on their own to try to find solutions. Let students know they will hear others' solutions during the class discussion that follows.

The largest box allowed by the model is 12 cm in each dimension. Thus, students cannot build a 1 cm × 1 cm × 24 cm box.

9. If students will save their work, have them do so now.

Project the sketch. Expect to spend about 30 minutes.

SUMMARIZE

10. Gather the class. Students should have their worksheets. ***We've been calling our Sketchpad model a box. Mathematicians call it a rectangular prism.*** Review the properties of rectangular prisms.
11. Tell the class that the amount of space inside a three-dimensional shape is called the *volume* of the shape. Another way to say this is that volume tells the amount of space a three-dimensional shape takes up. Students have found the volume of boxes by finding the number of cubes that fill the boxes.

Have students practice using the terms *volume* and *cubic centimeter* throughout this discussion.

Go to page "Build a Box" and show one cube. (Students can examine centimeter cubes as well if they have them.) ***The unit of measure you have been using is called a* cubic centimeter. *Each edge of this cube is 1 centimeter in length. Each face has an area of one square centimeter.*** Drag points to build several small boxes, and ask students to tell the volume of each box expressed in cubic centimeters.

12. Facilitate discussion of methods students have found for determining the volume of rectangular prisms. Begin by saying, I'm thinking of a box that has 2 rows of 10 cubes in the bottom layer, and 5 layers. Write this information on the board.

    How can I figure out the volume of the box? The discussion should bring out the ideas expressed in the student responses that follow.

    First, you have to figure out the number of cubes in a layer. Then multiply that by the number of layers. The box has 2 × 10 cubes in a layer; that's 20. Times 5 is 100. So the volume of the box is 100 cubic centimeters. This is the same as the way we made boxes with the model. First, you make the bottom layer; then you make more layers that have the same number of cubes.

    To find the number of cubes in a layer, just find the area of the bottom of the box. I think about each cube in the bottom layer sitting <u>on</u> one square centimeter, so the number of square centimeters in the area of the bottom of the box is equal to the number of cubic centimeters in the layer. The area of the bottom of your box is 2 × 10. Multiply that by the height, 5, which tells you the number of layers.

    Dragging point L creates the length of a box and dragging point W creates the width of a box. We can find the area of the bottom layer by multiplying the length (L) times the width (W). H is the height of the box. After we find the area of the bottom, we multiply by the height.

13. Discuss the Explore More problems, worksheet steps 7–9. In step 9, students may use this reasoning: The largest amount of cubes Roberto can use is 90 (the largest multiple of 5 that is less than 93); 20 layers would require 100 cubes, so 90 cubes will fill two layers of 5 fewer, or 18 layers; and 3 cubes will be left over.

EXTEND

1. ***What other questions about building and filling boxes occurred to you? What have you wondered about?*** Encourage student inquiry. Here are sample student queries.

   Why don't the large boxes look very realistic?

   Are there patterns that we can use to predict how many different rectangular prisms there are with a certain volume?

   Do boxes with the same volume have the same surface area? If they don't, what's the biggest surface area a box can have for the volume it has?

2. Have students write problems about building or filling boxes. One interesting problem type to introduce if students don't is this: ***A box is filled with exactly 30 cubes. The height of the box is 3 cm. What is the length and width of the box?*** Students should have no trouble determining that there are 10 cubes in a layer. They may be intrigued by the discovery that it is not possible to determine the box's length and width. Possible dimensions, in centimeters, are 1×10, 10×1, 2×5, and 5×2.

ANSWERS

2. Several answers are possible. Two solutions are 2 layers of 4 cubes, and 1 layer of 8 cubes.

3. Many answers are possible. Two solutions are 3 layers of 4 cubes, and 2 layers of 6 cubes.

4. Because 11 is a prime number, no set of numbers other than 11, 1, and 1 can be used as the dimensions of the box. The two possible solutions are 1 layer of 11 cubes, and 11 layers of 1 cube.

5. Many answers are possible. Two solutions are 6 layers of 4 cubes, and 4 layers of 6 cubes.

6. Many answers are possible. Two solutions are 6 layers of 6 cubes, and 4 layers of 9 cubes.

7. 7 layers

8. This is impossible. No even number is a factor of 27.

9. 18 layers

Stack It Up

Name:

Build boxes by making layers of cubes.

EXPLORE

1. Open **Stack It Up.gsp.** Go to page "Build a Box."
   You will drag points *L*, *W*, and *H* to make boxes.

2. Make a box with exactly 8 cubes. Record how many:
   cubes in each layer ___________
   layers ___________

3. Make a box with exactly 12 cubes. Record how many:
   cubes in each layer ___________
   layers ___________

4. Make a box with exactly 11 cubes. Record how many:
   cubes in each layer ___________
   layers ___________

Go to page "Two Boxes." Now you will build two different boxes, each with the same number of cubes.

5. Make two boxes, each with exactly 24 cubes. Record in the table.

	Cubes in a Layer	**Layers**	**Total Cubes**
Box 1			
Box 2			

6. Make two boxes, each with exactly 36 cubes. Record in the table.

	Cubes in a Layer	**Layers**	**Total Cubes**
Box 1			
Box 2			

EXPLORE MORE

Answer these questions. Tell about your thinking. If you want, go back to page "Build a Box" and use the model.

7. Christa used exactly 42 cubes to build a box. Each layer had 6 cubes. How many layers did the box have? ___________

8. Marta is making a box with exactly 27 cubes. She wants to use an even number of cubes in each layer. How can she do that?

____

____

____

____

9. Roberto has 93 cubes. How many layers high can he build the box if 5 cubes fill each layer?

____

____

____

____

Perfect Packages: Surface Area and Volume

INTRODUCE

Project the sketch for viewing by the class. Expect to spend about 10 minutes.

1. Open **Perfect Packages.gsp.** Go to page "Box and Net."
2. Explain, ***Today you'll look at two different ways of describing the size of a package that is a rectangular prism. On the left side of the sketch you see a package that has a volume of 6 cubic units. On the right side this box has been unfolded into a net and the area of the net corresponds to the surface area of the package. What is the surface area of this package?*** Allow students to propose answers and encourage them to describe how they calculated their answers. It may be useful to point to the different-colored rectangles on the net and relate them to the sides of the package. ***As you can see, the surface area and the volume are not usually the same. Sometimes, depending on the purpose of the package, people want to make a box that uses as little surface area as possible while having a certain volume. Can you think of an example of when this might happen?***
3. Show students how to drag the sliders to change the length, width, and height. Ask, ***Can someone tell me how I could make a package that has the same volume as this one, but a different shape?*** Use the sketch to illustrate students' examples. These might include $1 \times 1 \times 6$. Ask students whether a $3 \times 1 \times 2$ package should be considered to have the same shape as a $2 \times 3 \times 1$ package. Opinions may differ, but let students know that this activity assumes they are the same shape. Ask, ***Do you think that all the packages with a volume of 6 cubic units have the same surface area? This is the first question you'll be exploring today.***

DEVELOP

Expect students at computers to spend about 25 minutes.

4. Assign students to computers and tell them where to locate **Perfect Packages.gsp.** Distribute the worksheet. Tell students to work through step 7 and do the Explore More if they have time.
5. Let pairs work at their own pace. As you circulate, here are some things to notice.
   - Help students find shorter ways than counting to calculate the surface area of the box.
   - Help students develop strategies to make sure they have found all possible ways of creating a box with a given volume. This might

involve making a table or listing all the factors of the value of the volume.

- Encourage students who finish quickly to investigate dimensions that are not whole number values (in the Explore More).

SUMMARIZE

Project the sketch. Expect to spend about 10 minutes.

6. Open **Perfect Packages.gsp** and use it to support the discussion. Invite students to share their strategies for finding the surface area of a package. If students bring it up, you might add to the projected sketch the surface area calculation described in the Extend section.
7. Ask students to talk about which packages had the least and greatest surface areas for a given volume. Students should notice that the least surface area occurs when the dimensions of the package are as equal as possible (that is, when the package is as cubic as possible), and that the greatest surface area occurs when two of the dimensions are very small and the third dimension is very large (that is, when the package is as long and thin as possible). Encourage students to consider dimensions that are not whole numbers. In this case the least surface area will occur when the dimensions are all the cube root of the volume.

EXTEND

1. Use Sketchpad's calculator to write a general equation for the surface area of the package (such as $2*length + 2*width + 2*height$, or $2*(length + height + width)$). Change the value of one of the dimensions and investigate how both the volume and the surface area change.
2. Discuss how the relationship between volume and surface area might change for prisms that don't have right angles, or for other solids. If students already know that a circle gives the maximum area for a given perimeter, you can discuss how a sphere gives maximum volume for a given surface area.

ANSWERS

1. Surface area = 22 square units
2. 1 × 1 × 6 box: volume = 6 cubic units; surface area = 26 square units

   1 × 1 × 5 box: volume = 5 cubic units; surface area = 22 square units
3. The values for length, width, and height are interchangeable.

   2 × 3 × 4 arrangement: surface area = 52 square units

   2 × 2 × 6 arrangement: surface area = 56 square units

   1 × 4 × 6 arrangement: surface area = 68 square units

   1 × 3 × 8 arrangement: surface area = 70 square units

   1 × 2 × 12 arrangement: surface area = 76 square units

   1 × 1 × 24 arrangement: surface area = 98 square units
4. The 2 × 3 × 4 arrangement has the least surface area. The 1 × 1 × 24 arrangement has the greatest surface area. (On page "Decimal Values," using approximately 2.9 units for each dimension will produce the least surface area.)
5. a. 2 × 2 × 2 (surface area = 24 square units)
   b. 2 × 2 × 3 (surface area = 32 square units)
   c. 3 × 3 × 3 (surface area = 54 square units)
   d. 3 × 4 × 4 (surface area = 80 square units)
6. a. 1 × 1 × 8 (surface area = 34 square units)
   b. 1 × 1 × 12 (surface area = 50 square units)
   c. 1 × 1 × 27 (surface area = 110 square units)
   d. 1 × 1 × 48 (surface area = 194 square units)
7. To find the least surface area, make the dimensions as equal as possible (so the package is as close to a cube as possible). To find the greatest surface area, make two of the dimensions equal to 1, and the third dimension equal to the volume (to make as long and thin a package as possible).
8. The package with the least surface area for 5b (volume = 12) is when each dimension is approximately 2.3 units. The package with the least surface area for 5d (volume = 48) is when each dimension is approximately 3.6 units.

The greatest surface area for each volume is when two dimensions are 0.1 unit and the third dimension is very large, but these no longer represent realistic packages.

9. Answers will vary. Sample answers: $1 \times 1 \times 5.5$ (volume = 5.5 cubic units); $1 \times 1.6 \times 4$ (volume = 6.4 cubic units); $2 \times 2 \times 2$ (volume = 8 cubic units)

Perfect Packages

Name:

In this activity you'll investigate how changing the dimensions of a package (a rectangular prism) affects volume and surface area. Of course, you'll be most interested in the biggest and smallest packages you can make!

EXPLORE

1. Open **Perfect Packages.gsp** and go to page "Box and Net."

   If needed, change the dimensions to match those pictured at right, so the volume is 6 cubic units. What is the surface area?

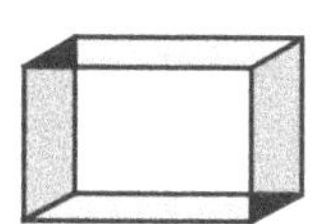

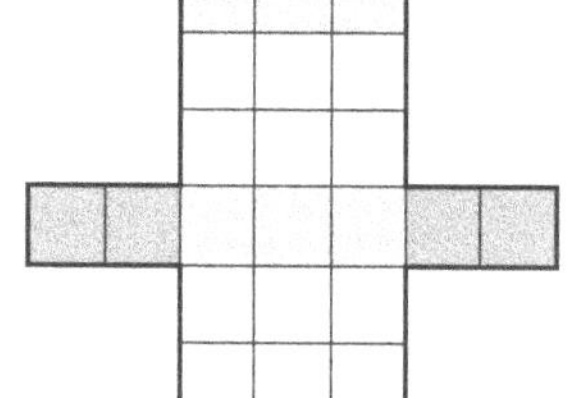

2. Can you find a package that has the same volume, but a different surface area? Can you find a package with the same surface area, but different volume? Drag the length, width, and height sliders to see whether you can come up with examples of such packages. Describe what you found.

3. Find all the ways 24 cubes can be arranged into a package. For each arrangement, write the dimensions and surface area in the table.

Length	Width	Height	Volume	Surface Area
			24 cubic units	
			24 cubic units	
			24 cubic units	
			24 cubic units	
			24 cubic units	
			24 cubic units	
			24 cubic units	
			24 cubic units	

Perfect Packages

continued

4. Which of your arrangements has the least surface area? Which has the greatest?

5. For each given volume, which package has the least surface area?
   a. Volume = 8 cubic units
   b. Volume = 12 cubic units
   c. Volume = 27 cubic units
   d. Volume = 48 cubic units

6. For each given volume, which package has the greatest surface area?
   a. Volume = 8 cubic units
   b. Volume = 12 cubic units
   c. Volume = 27 cubic units
   d. Volume = 48 cubic units

7. For any given volume, describe how you would find the packages with the least and greatest surface areas.

EXPLORE MORE

8. Go to page "Decimal Values," where you can adjust the dimensions to the nearest tenth of a unit. See whether you can find packages whose surface areas are less than those you found in step 5, or greater than those you found in step 6. Describe what you found.

9. Suppose you have a surface area of 24 square units. Can you make packages with different volumes? If so, provide two possibilities.

Fly on the Ceiling: Coordinate Systems

ACTIVITY NOTES

INTRODUCE

Project the sketch for viewing by the class. Expect to spend about 10 minutes.

1. Open **Fly on the Ceiling.gsp.** Go to page "Ceiling." Enlarge the document window so it fills most of the screen.
2. Explain, ***Today you are going to use Sketchpad to explore a coordinate plane as mathematician and philosopher René Descartes first saw it—as a fly crawling on a ceiling.*** Recount the story about Descartes lying in his bed and realizing that he could describe a fly's position on the ceiling by using two numbers that described its distance from each of two walls.

Descartes is perhaps better known as a philosopher than as a mathematician. His most famous quote is *Cogito ergo sum*—"I think, therefore I am."

3. Model worksheet steps 1 and 2 for the class. Here are some tips.
   - In worksheet step 1, model dragging the point. ***What happens to the measurements on the sketch and in the table?*** [The measurements change as the point moves.] ***The measurements are called*** **coordinates.** ***A coordinate is one of the two numbers that locates a point on a coordinate plane. The coordinate plane in this case is the ceiling.***
   - Review with students that it takes two coordinates to locate a point on a coordinate plane. ***Suppose I tell you the fly is 2 meters from the left side of the ceiling. Can you tell me exactly where the fly is?*** [No, students will only be able to state that the fly is somewhere 2 meters from the left side.] ***What other information do you need to find the fly's location?*** [How far it is from the bottom (or top) side]
   - Make sure students understand that the numbers in the table correspond to the metric measurements from the two walls to the point. Drag the point to (2, 1). ***How far is the fly from the left side of the ceiling?*** [2 meters] ***How far is the fly from the bottom side of the ceiling?*** [1 meter] ***What do the numbers in the table tell you?*** [The x-value is the distance from the left side; the y-value is the distance from the bottom side.]
   - Now drag the point to (1.3, 2.4). ***How far is the fly from the left side of the ceiling now?*** [1.3 meters] ***From the bottom?*** [2.4 meters] Be sure students understand that the measurements are in tenths of a meter. ***How are the distances measured?*** [In tenths of a meter]
   - In worksheet step 2, the terms x- and y-coordinates are introduced as names for the two coordinates. Tell students that (0, 0) is known as the *origin* of a coordinate plane. ***The coordinates tell the distance***

from the origin. The x-coordinate tells the distance left or right from the origin. The y-coordinate tells the distance up or down from the origin. In which corner is the origin? [Bottom-left corner] Drag the point to the bottom-left corner to show students the origin.

- Now drag the point to the top-right corner. ***How far is the fly from the left side of the ceiling?*** [6 meters] ***How far is the fly from the bottom side of the ceiling?*** [4 meters] ***How could we name the location of this point?*** Students may suggest (4, 6) or (6, 4). ***There is a standard way of referring to points, so there isn't any confusion about the location. The left and right, or horizontal, distance from the origin is listed first. This is the x-coordinate. The up and down, or vertical, distance from the origin is listed second. This is the y-coordinate.*** Explain that the two coordinates for the point in the top-right corner are (6, 4).

Students can double-click the table with the **Arrow** tool while holding down the Shift key to remove the most recently added value.

- Demonstrate how to double-click the table to enter a measurement.

4. ***Now you will explore Descartes' coordinate plane on your own. As you work, think about how you can use coordinates to find locations and to determine measurements.***
5. If you want students to save their work, demonstrate choosing **File | Save As,** and let them know how to name and where to save their files.

DEVELOP

Expect students at computers to spend about 35 minutes.

6. Assign students to computers and tell them where to locate **Fly on the Ceiling.gsp.** Distribute the worksheet. Tell students to work through step 22 and do the Explore More if they have time.
7. Let pairs work at their own pace. Encourage students to ask their neighbors for help if they are having difficulty using Sketchpad. As you circulate, here are some things to notice.
   - In worksheet step 3, some students may not drag the point completely into the corner. Tell students to drag the point as far as possible in both directions. The coordinates will be whole numbers.
   - In worksheet step 4, listen to students as they discuss the answer to this problem. You may hear a range of comments that can inform your teaching. Here are some sample student responses.

     The two right corners are both 6 meters from the left side of the ceiling. The two top corners are both 4 meters from the bottom side of the ceiling. That makes the dimensions 6 meters by 4 meters.

Look at the coordinates of the top-right corner, (6, 4). The 6 tells us that the ceiling is 6 meters long. The 4 tells us the ceiling is 4 meters wide.

If we drag the point along the bottom edge toward the far right, we can read the number after the arrow. That is the length. Then let's drag the point up along the right side to the top-right corner. That number is the width.

- In worksheet steps 5 and 6, if students have trouble finding the center, have them think about one dimension at a time. ***How would you find the center of the horizontal distance? How would you find the center of the vertical distance? What are the coordinates of a point that is at the center of both?***

Have students drag the fly into a corner so they can see the points more easily.

- In worksheet steps 8–10, students should recognize the vertices of each shape. Encourage students who say, for example, just "a square" to write more in their descriptions. ***What else can you say about the shape? For example, can you find the length of each side?***

Students can select and drag the measurements to move them next to each point.

- In worksheet step 11, for students who have trouble identifying the points, ask them to recall what the properties of a square are. ***What are the properties of a square? How can you use these properties to help you find the coordinates of the missing vertices?***
- In worksheet steps 12 and 13, remind students of the properties of a right triangle and an isosceles triangle, if needed. ***How would you describe a right triangle?*** [A triangle with a 90° angle] ***How can you plot a point to create a 90° angle?*** [The point must create two perpendicular sides when the points are connected to form the triangle.] Continue questioning in a similar manner for an isosceles triangle.
- In worksheet step 14, students should have fun swatting Descartes' flies. Watch for students who reverse the *x*- and *y*-coordinates. ***Do* (0.7, 3.0) *and* (3.0, 0.7) *name the same point? Explain.***

If you do this activity over two days, start the second day here.

- In worksheet steps 15 and 16, students explore the four quadrants on the coordinate plane. By dragging the fly around, they should notice that the *x*-value is negative when the fly is to the left of the origin and the *y*-value is negative when the fly is below the origin. Be sure students understand that the light is the origin. ***What are the coordinates of the light?*** [(0, 0)]

- In worksheet steps 17–19, students will discover that the position of the origin does not make a difference when using coordinates to measure distances. This may be confusing to some students. Stress that all measurements are positive values and that the coordinates tell how far a point is from the origin. ***Can the room's length or width have a negative value?*** [No] ***What does a negative x-value tell you?*** [How far a point is to the left of the origin] ***What does a negative y-value tell you?*** [How far a point is below the origin]
- In worksheet step 21, students find the coordinates for two missing vertices of a rectangle. Ask students to think about the relationship between the missing vertices and the two known vertices. ***How can you figure out where the missing vertices go? What are their positions in relation to the known vertices?*** If students need help, have them press *Show Coordinate System* to see the coordinate plane.
- If students have time for the Explore More, they will have fun plotting points to create designs and puzzles for their classmates to solve.

8. If students will save their work, remind them where to save it now.

SUMMARIZE

Project the sketch. Expect to spend about 15 minutes.

9. Gather the class. Students should have their worksheets with them. Open **Fly on the Ceiling.gsp** and go to page "Ceiling." Discuss worksheet steps 2 and 3. ***We are looking at Descartes' ceiling. The fly moves to the bottom-left corner. How can you use coordinates to tell where the fly is?*** [(0, 0)] Drag the point to that location to verify the coordinates. Double-click the table to enter the measurements. Continue in a similar way for each of the remaining three corners.
10. ***How did you determine the length and the width of the room?*** Have volunteers explain their methods. Make sure students understand that they need to find the range of the *x*- and *y*-coordinates.
11. Discuss how students found the answers to worksheet steps 6 and 7. Students' responses may vary. Here are some sample solutions to worksheet step 6.

    We know the length of the ceiling is 6 meters, so the center of the length would be half of that, or 3 meters. The width of the ceiling is 4 meters. Half of that is 2 meters. We figured the center would be 3 meters over and 2 meters up, at (3, 2).

The x-coordinates for the two bottom corners are 0 and 6. To find the center, we added them and divided by 2 to get 3. The y-coordinates for the bottom and top corners are 0 and 4. To find the center, we added them and divided by 2 to get 2. Our center point is (3, 2).

12. Have students discuss their answers to worksheet steps 9 and 10. Depending on the level of your students, you may wish to discuss finding the perimeter and the areas of the shapes. ***What is the formula for the area of a square?*** [s^2] ***What is the length of one side? How do you know?***

13. Review worksheet steps 11–13, having students share their strategies for finding the coordinates of the missing points.

14. Discuss the four quadrants of a coordinate plane and how the values of the coordinates change in each one. Check for understanding. ***In which coordinate would you find the following points?***
    - point A (−4, 3) [Quadrant II]
    - point B (−4, −3) [Quadrant III]
    - point C (4, 3) [Quadrant I]
    - point D (4, −3) [Quadrant IV]

15. In worksheet steps 17 and 18, discuss how students found the dimensions of the room and how they are the same as in worksheet step 4. ***Suppose I move the origin to the top-right corner. Do the dimensions of the room change? Explain.*** [No, the dimensions are independent of the origin.]

16. In worksheet step 21, have students share their strategies for finding the missing vertices of the rectangle.

17. If time permits, discuss the Explore More. Students can share their designs with the class.

18. You may wish to have students respond individually in writing to this prompt. ***A square has side lengths of 3 meters. Explain where the square is for each description of its vertices.***
    - ***All vertices have positive coordinates.*** [Quadrant I]
    - ***All vertices have negative coordinates.*** [Quadrant III]
    - ***Two vertices have negative x- and y-values and two vertices have positive x-values and negative y-values.*** [Quadrants III and IV]

EXTEND

1. Ask students to find other types of triangles on page "Triangles" (worksheet steps 12 and 13) such as isosceles right triangles or equilateral triangles. Placing the third vertex at (3, 3) or (5, 3) forms an isosceles right triangle. Placing the third vertex at (4.0, 2.7) approximates an equilateral triangle.
2. Explore missing vertices of a rectangle further. ***If the opposite vertices of a rectangle have coordinates (a, b) and (c, d), name one possible set of coordinates for the other two vertices using only a, b, c, and d.*** [(a, d) and (c, b)]

ANSWERS

4. The room is 6 meters long and 4 meters wide. These are the ranges of the x- and y-coordinates.
6. The coordinates of the center of the rectangle are (3, 2). These are half the values of the ranges of the x- and y-coordinates.
7. The coordinates of the point are (4, 1).
9. The figure is a square with side length 2 m. Its perimeter is 8 m and its area is 4 m^2.
10. The figure is an isosceles trapezoid. The bases are 2 m and 4 m. Depending on the background of your students, they might also calculate that the two other sides are each $\sqrt{5}$ m, the perimeter is $(6 + 2\sqrt{5})$ m, and the area is 6 m^2.
11. The missing coordinates are (3.0, 1.0) and (5.5, 3.5).
12. Answers will vary. There are many points that can be used to make a right triangle. Most of them have an x-coordinate of either 3 or 5. Possible solutions using only integers: (3, 2), (3, 3), (3, 4), (5, 2), (5, 3), (5, 4)
13. Answers will vary. There are many points that can be used to make an isosceles triangle. All of them either have an x-coordinate of 4 (the average of the two given x-coordinates) or have the same x-coordinate as one of the given points and a y-coordinate of 3. Possible solutions using only integers: (4, 2), (4, 3), (4, 4), (3, 3), (5, 3)

14. The constellation is the Big Dipper, one of the most distinctive constellations in the northern sky.
16. The x-coordinate has a negative value when it is left of the origin. The y-coordinate has a negative value when it is below the origin.
18. The room is 6 meters long and 4 meters wide. These are the ranges of the x- and y-coordinates.
19. The dimensions for the room are the same. The size of the room is independent of the position of the origin.
20. In Quadrant I, both coordinates are positive.

    In Quadrant II, x is negative and y is positive.

    In Quadrant III, both coordinates are negative.

    In Quadrant IV, x is positive and y is negative.
21. $(-2.4, 1.6)$ and $(1.0, -1.0)$
22. The figure is a regular hexagon.
23. Designs will vary. Check students' work.
24. Designs will vary. Check students' work.

Fly on the Ceiling

Name:

One day, philosopher and mathematician René Descartes noticed a fly walking on the ceiling. Descartes realized that he could describe the fly's position on the ceiling by two numbers: its distance from each of two walls. Thus was born the *coordinate plane,* also called the *Cartesian coordinate system* after Descartes. You'll explore the coordinate plane in this activity.

EXPLORE

Open **Fly on the Ceiling.gsp** and go to page "Ceiling." You will see a model of the ceiling in Descartes' bedroom. The red point is the fly in the story.

1. Drag the fly and notice how the measurements change on the sketch and in the table to the right. These measurements are called *coordinates.* You can also move the fly in very small steps using the Arrow keys on the keyboard.

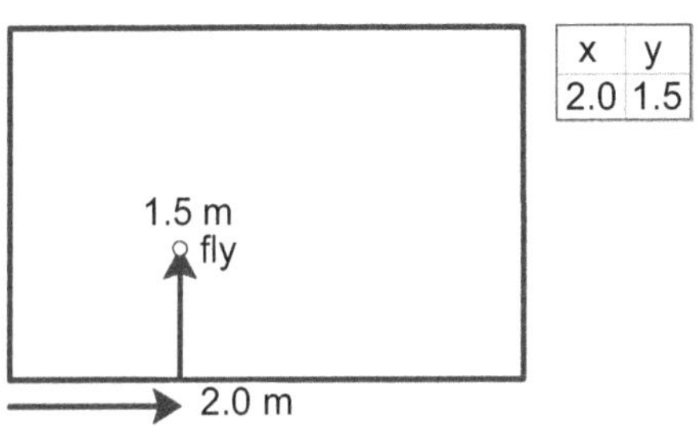

x	y
2.0	1.5

2. Drag the fly to the top-right corner. Double-click the table to enter the measurements of the *x*- and *y*-coordinates.

3. Drag the fly to the other three corners and enter their coordinates into the table. Compare your four pairs of measurements with the measurements of other students.

4. How long and wide is the room? Explain how you can figure this out from the coordinates of the four corners.

5. Drag the fly as close as possible to the center of the room.

6. What are the coordinates for the center point? Describe how you can figure this out using the coordinates of the four corners.

7. If the fly is resting $\frac{1}{4}$ from the bottom edge and $\frac{2}{3}$ from the left edge, what are its coordinates?

CONSTRUCT

Now you will plot points using their coordinates.

8. Plot these four points: (2, 1), (2, 3), (4, 1), and (4, 3). To plot points, choose **Graph | Plot Points.** For each point, enter its coordinates and click **Plot.** After you enter the last point, click **Done.**

9. Describe in detail the figure outlined by the points.

10. Go to page "Mystery Shape." The data in the table represent the last four locations that the fly rested on the ceiling, which happen to form a special shape. Drag the fly to each point given in the table. Can you visualize the shape? Now select the table and choose **Graph | Plot Table Data.** Then click Plot. Describe the figure outlined by the points.

11. Go to page "Missing Corners." The two points are opposite vertices of a square. What are the coordinates for the two missing vertices? Plot them. You can find the coordinates of a point by selecting it and choosing **Measure | Coordinates.**

12. Go to page "Triangles." The two points are two vertices of a triangle. Write down a pair of possible coordinates for the missing vertex so that the triangle is a right triangle. Plot the missing point. Then write down another pair of possible coordinates that would form a right triangle.

13. Delete the point you plotted in the last step. Write down a pair of possible coordinates for the missing vertex so that the triangle is an isosceles triangle. Plot the missing point. How is the *x*-coordinate of the missing vertex related to the *x*-coordinates of the original two points?

Go to page "Fly Swatter." René looked back up at his ceiling and noticed a bunch of flies, so he grabbed his fly swatter to get rid of them. The table lists the coordinates of each fly he swatted. When he looked back up at the ceiling, he noticed that the squashed flies resembled a well-known constellation.

14. Drag the fly to each point listed in the table and press *Fly Swatter.* Then plot the table data to check your work. What is the name of the constellation?

EXPLORE

So far you've seen how Descartes measured the coordinates of a fly by noting its distance from each of two walls. It's possible to use instead a single fixed point, such as a light hanging in the middle of the room, as a reference point.

This reference point is called the *origin* of a coordinate system. So far the origin has been the bottom-left corner, but now you'll use a different origin.

15. Go to page "Ceiling with Light." Notice how the position of the fly is now described by using horizontal and vertical measurements from the location of the light. The light is the origin of a new coordinate system. Drag the fly and notice how the measurements change on the sketch and in the table to the right.

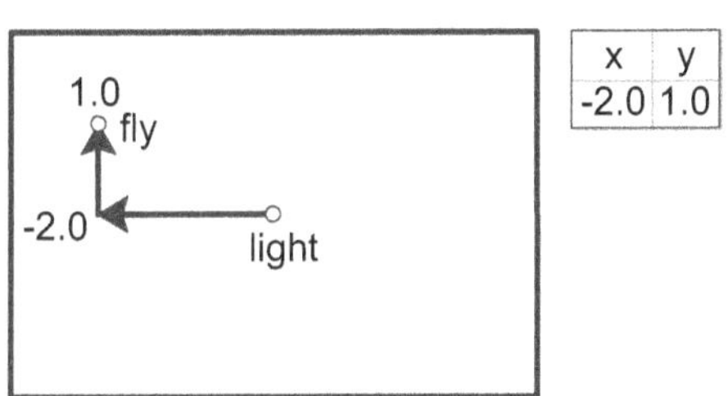

16. What happens to the *x*-coordinate when the fly is to the left of the light? And what happens to the *y*-coordinate when the fly is below the light?

17. Drag the fly to the top-right corner. Double-click the table to enter the measurements of the *x*- and *y*-coordinates. Then drag the fly to the other three corners and enter their coordinates into the table.

18. How long and wide is the room? Explain how you can figure this out from the coordinates of the four corners.

19. Did you get the same dimensions for the room in step 18 as you did in step 4 when the origin was at the bottom-left corner? Does the location of the origin make any difference when using coordinates to measure distances?

20. Press *Show Coordinate System.* The *x*- and *y*-axes divide the plane into four regions called quadrants, numbered I, II, III, and IV, as shown here. For each of the quadrants, state a general rule about the signs of the coordinates of a point in that quadrant.

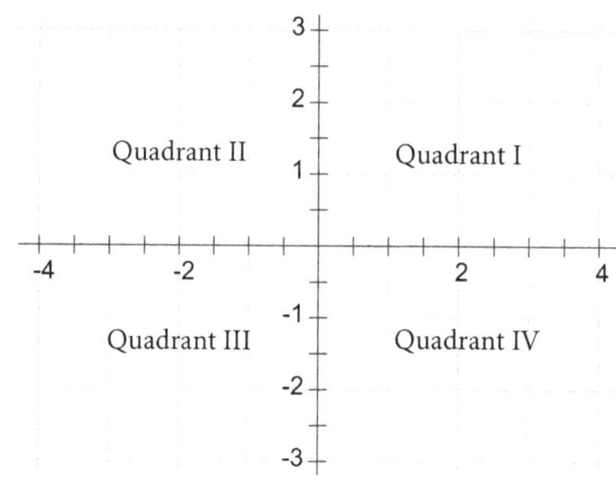

21. Go to page "Incomplete Rectangle." The two points are opposite vertices of a rectangle. What are the coordinates for the two missing vertices? Plot them.

22. Now go to page "Hidden Shape." Look at the table and try to visualize where the points will appear when you plot them. Drag the fly to help you. Can you visualize the hidden shape? Then plot the table data. Describe the figure outlined by the points.

EXPLORE MORE

23. Go to page "Target." Double-click on the values for *x* and *y* and change them to move the target. Then press *Go to Target* to move the arrow that leaves a trace. Once you get the hang of it, choose **Display | Erase Traces** to start over and create a design of your own.

24. Go to page "Make Your Own" and create your own design out of points. Drag the fly to each point, double-click the table to add the coordinates to the table, and then delete the point. Have a classmate try to figure out what your design is by looking at the table and then check by plotting the table data.

If you are only doing a short demonstration after students have already derived the formula for circle area, show the model, and then begin your discussion with step 6.

INTRODUCE

Project the sketch for viewing by the class. Expect to spend about 20 minutes.

1. Open **Wavy Parallelogram.gsp** and go to page "Circle Sectors." Ask students to study and then describe the two representations. ***How are the circle and this "wavy parallelogram" shape related?*** [Each is made up of the same number of circle sectors of the same size.]
2. Give students plenty of time to study the representations, and then begin discussion. Students are likely to recognize that the sectors of the circle have been rearranged in a way that approximates a parallelogram. (Inside the wavy parallelogram is the outline of an actual parallelogram.) ***What can you say about the area of these figures?*** Elicit the idea that the area of the two figures must be the same.
3. Ask the following two questions, and then use a think-pair-share approach to facilitate students' making sense themselves of the relationships between the parts of the wavy parallelogram and the parts of the circle.

   As you trace along the wavy base, say, ***Look at the wavy base of the wavy parallelogram. How is it related to the circle?*** [It's half the circumference.]

   How is the height of the wavy parallelogram related to the circle? [It's the radius.]

   The discussion should bring out the ideas that the base of the wavy parallelogram is half the circle's circumference and that the height is the circle's radius.

   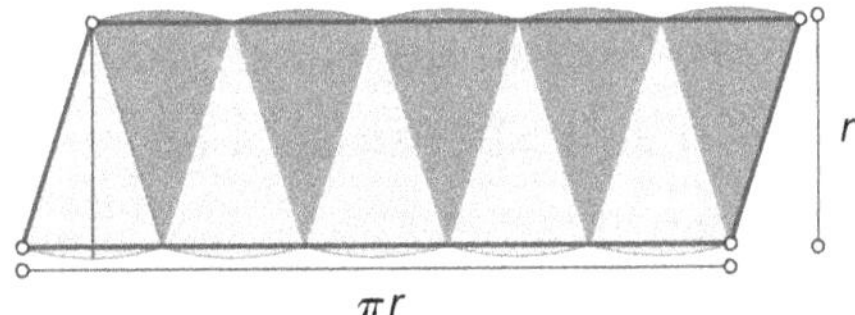

4. Drag the slider slowly, pausing at 6, 8, and 10 and stopping at 12. Students will observe that as the number of sectors increases, the wavy parallelogram more closely approximates a parallelogram. ***You observed that the area of the circle and the area of the wavy parallelogram are the same. How can you use what you know about parallelogram area***

to find the area of the circle? Work together to figure this out. Allow plenty of time for students to work in pairs or small groups and struggle with this question. If you need to prompt some students, ask questions such as these.

How can knowing the formula for the area of a parallelogram help you? Make sure students recall the formula.

The base of the wavy parallelogram is half the circumference of the circle. How can you use what you know about the formula for the circumference of a circle, $C = \pi d$, or $C = 2\pi r$?

5. When students are ready, bring the class together again and invite students to share their thinking. ***Let's hear your ideas about finding the area of the circle.*** The discussion should bring out the following ideas.
   - Using r for radius of the circle, the circumference of a circle is equal to $2\pi r$. Because the base of the parallelogram is half of the circumference, it is equal to πr.
   - The height of the parallelogram is equal to the radius of the circle, r.
   - The formula for area of a parallelogram is $A = b \cdot h$.
   - We can substitute πr for the base, and r for the height.
   - Multiplying base times height in the wavy parallelogram gives $\pi r \cdot r$, or πr^2.
   - Let students know that this new formula is a formula for circle area.

DEVELOP

Expect to spend about 15 minutes. Continue to project the sketch.

6. ***Are you convinced that using the formula for the area of a regular parallelogram in the way we did works for a wavy parallelogram? Were any of you concerned that a wavy parallelogram isn't a "real" parallelogram?*** Give students a moment to express any uncertainty. Acknowledge their responses and express respect for their thoughtfulness.

   Sketchpad gives us a way to look a little deeper into why your formula works. We can drag the slider more and that will increase the number of sectors. How will increasing the number of sectors affect how wavy the parallelogram is? Students are likely to intuit correctly that fewer sectors make wavier parallelograms and that more sectors result in a straighter base. Drag the slider all the way to 18 to illustrate this.

7. Drag the slider back to 6 and direct students' attention to the actual parallelogram constructed within the wavy parallelogram. ***Why is the area of the actual parallelogram less than the area of the circle?*** Students should observe that wavy parts of the circle extend outside the parallelogram because the radius of the circle is greater than the height of the parallelogram.

8. ***What do you think will happen to the areas of the parallelogram and circle as I increase the number of sectors again? Why?*** Students predict that the areas will get closer to one another—specifically, the area of the parallelogram will increase while the area of the circle stays constant. The parallelogram area increases because the height gets closer to the radius of the circle. Drag the slider again, slowly, and discuss students' observations and thinking. Follow up with three more questions.

   What do you notice about the base of the actual parallelogram compared to the base of the wavy one as the number of sectors increases? [The base of the wavy parallelogram more closely approximates the base of the actual parallelogram.] ***What if we had 100 sectors? A 1000?***

   What do you notice about the sides of the parallelogram as the number of sectors increases? [The sides make a larger angle with the base as the number of sectors increases.] ***What would the sides be like with a very large number of sectors?***

   Compare the height of the actual parallelogram and the height of the wavy one. [The height of the wavy parallelogram more closely approximates the height of the actual parallelogram as the number of sectors increases.] ***How would those heights compare if there were a very large number of sectors?***

SUMMARIZE

Expect to spend about 10 minutes. Continue to project the sketch.

9. ***How does the Sketchpad demonstration help confirm that we really can use parallelogram area to find the area of a circle?***

   The more sectors the circle is cut into, the more closely the circle approximates a parallelogram.

   If we could cut a circle into a very large number of sectors, we would get a parallelogram.

If we could cut a circle into infinitely many sectors, we would get a rectangle.

Let students use their own form of inductive reasoning to extend their thinking to a very large number of sectors.

There is no limit to how close we can get to a parallelogram.

We get so close that the parallelogram becomes a rectangle.

We can derive the formula for the area of a circle from the formula for the area of a rectangle.

10. Suggest that the class write an explanation for deriving the formula for the area of a circle. You might act as scribe, recording the explanation and asking for clarification as needed in order to encourage clear communication. Alternatively, have students write individually in response to the following prompt.

    How were you able to use what you know about parallelogram area to come up with a formula for circle area?

EXTEND

1. Show page "Circle in Square," a circle and a square around it whose side midpoints just touch the circle, as shown. ***What is the area of this square?*** [$A = 4r^2$] You might challenge students to construct a Sketchpad circle and square that will always have this relationship.

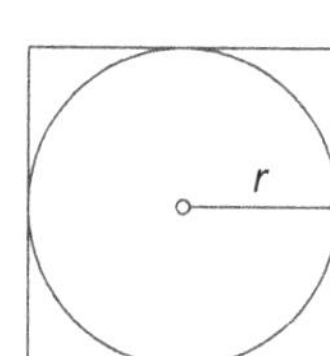

Making Means: Data Distribution and Averages

ACTIVITY NOTES

INTRODUCE

Project the sketch for viewing by the class. Expect to spend about 5 minutes.

1. Open **Making Means.gsp** and go to page "Line Plot."
2. Explain, ***Today you're going to use Sketchpad to create data sets that have specific mean values. Usually, you start with a given data set and have to calculate the mean. But now you'll work in the reverse direction. If you're given a certain mean value, can you think of how many data sets it might represent? How big would the data set have to be? How might the different values in the data set be distributed? Before you begin, I'll demonstrate how to create a data set using the sketch.***
3. Drag the data point from 7 to 6. Students should see that the mean changes as you do this. Drag a new data point to the number line and ask the students where to place the data point in order to make the mean smaller, bigger, and the same.
   - Show that data points can be stacked.
   - Show students that dragging a data point back to the left bank removes it from the set.
   - Press *Reset* to return to the initial configuration of the sketch.

DEVELOP

Expect students at computers to spend about 25 minutes.

4. Assign students to computers and tell them where to find **Making Means.gsp.** Distribute the worksheet. Tell students to work through step 16, and do the Explore More if they have time. You might consider convening the class after most students have completed worksheet step 11 and discussing students' predictions before asking students to continue on with the remaining questions.
5. Let pairs work at their own pace. As you circulate, here are some things to notice.
   - For questions where students are asked to provide more than one example, encourage them to construct examples that are different. For instance, {3, 4, 5}, {2, 4, 6}, and {1, 4, 7} are similar examples of data sets with means of 4.00. However, {0, 6, 6} is quite different because it neither includes 4 nor has distinct values.
   - Encouraging students to see what their sets of data have in common might help them construct different data sets. (Sometimes it's hard for

students to think outside a particular strategy they have developed to generate new data sets).

- For worksheet step 10, encourage students to write down things they know for sure and things they don't know for sure. Sometimes the things you don't know for sure can be just as important!
- In talking about the prediction for worksheet step 11, ask students to explain their predictions. Students can either use examples or describe how the sum of the numbers and total number of data points will change when adding a new data point.

6. If you decide to convene the class once most students have completed worksheet step 11, follow steps 7–10 in the Summarize section before having pairs return to their computers.

SUMMARIZE

Project the sketch. Expect to spend about 15 minutes.

7. Gather the class. Students should have their worksheets with them. Open **Making Means.gsp** and use it to support the class discussion.
8. Ask students to offer solutions to worksheet step 4, and write them down. After you have written down several, ask students for examples that are different from the ones written on the board. Help the students notice that not all data sets with a mean of 5 will be "balanced" around the mean or will include the value 5.
9. Now discuss the question in worksheet step 9, and ask students to talk about their strategies. Some will describe how they constructed their data sets so that the values always summed to 20. Some might also have considered the data points in pairs, and thought about summing a given pair to 10. If students are having difficulty using the sum to help them generate new data points, illustrate how they might do this using fewer data points and a different mean.
10. Students have now worked quite a bit on constructing different sets of data. Using the sketch, drag five data points to these locations: 2, 3, 4, 5, and 6. ***How does the mean change as more data points are added? Can the mean stay the same? Can it increase? Decrease? How do you know?*** Ask for volunteers to explain their reasoning. Then drag a new data point to 4, and drag it slowly toward the right so that the students see the mean value continuously increasing. Return it to 4, and ask a

student to describe what will happen if the value moves toward the left. This should help students see that the mean must change unless the new data point has the same value as the previous mean.

11. Continue on to worksheet step 14. Ask for volunteers to describe their answers. Some will have found a way of getting a mean of 10 by dragging the square off the screen. For those who stayed within the bounds, ask them to explain why the mean could never be 10. They may simply want to show you that they placed all of the remaining data points on 10, and still the mean was less than 10. They should be able to explain that at least one data value would have to be greater than 10 for the mean to be 10.00 given the existing data points. To provide some extra challenge: ***What if we placed three of the four remaining data points at 10? Then what value would the final data point have to be in order to make a mean of 10?*** [46]

12. For worksheet step 15, ask students to share some of the data points they had. Again, make sure to ask for different data sets. ***So, given these data sets, what can we say about a family of 8 that has a mean height of 4 feet?*** Encourage students to take into account the heights that the family members could realistically have.

EXTEND

1. ***When do you think that the mean would be a good measure to use to describe a set of data?*** Encourage students to give specific examples. They may have already encountered the mean in the context of school grades. Usually, what's important is not so much the kind of data that is being collected (grades), but what we want to know about the data. So, if we want to know whether a test was too hard for most students, then the mean is helpful because it is sensitive to extreme values (in this case, to very low test scores). However, the mean might not be a very good measure if we don't care about extreme values (maybe some students were away, so we don't want their marks of 0 to affect the measure).

2. Have pairs work together with the sketch. Introduce the idea of the median, if the students haven't already encountered it. Ask the students to find a few examples where the mean will equal the median for a given data set.

ANSWERS

3. Answers will vary. Sample answers: {3, 4, 6, 7}, {4, 4, 5, 7}, or {2, 6, 6, 6}

4. Answers will vary. See above.

6. {0, 9, 9} or {0, 8, 10}. Other answers are possible if data points beyond 10 are used.

9. Answers will vary. Sample answer: To keep a given mean when adding two data values, balance the two data points around the mean.

10. The data points can be distributed in many ways, and can even include extreme values such as 0 and 10. None of the data points has to be 4.00. However, the mean of all the data points must equal 4.00.

11. Answers will vary.

12. If the new data value is greater than 4, then the mean will increase. The larger its value, the more the mean will increase. Similarly, if it's less than 4, the mean will decrease, with smaller values decreasing it more.

14. Only if some of the four new data points have values more than 10. For the ten data values to have a mean of 10, the sum of the ten data points would have to be 100, which is 76 more than the sum of the given data values.

15. Answers will vary. The sum of the heights must be 32.

16. Because the question involves the heights of humans, it is unlikely that any person will be more than 7 feet tall and unlikely that any person will be less than 1 foot tall. In order to get a mean of 4, we know that either all of the heights are exactly 4, or at least some will be less than 4 feet tall, and some will be more.

17. The sum of the three points is 27, and the sum of the first two is 10, so the third point is 17.

Making Means

Name:

In this activity you'll investigate some properties of the mean and generate examples of data sets that have a given mean.

EXPLORE

1. Open **Making Means.gsp.** Go to page "Line Plot." Make sure you have one data point at 3 and one data point at 7. If not, press *Reset.*

2. Drag the two data points to new values without changing the mean value of 5.00. Find three ways of doing this.

3. Add two more data points so that the mean remains 5.00. Write down your new set of data points.

4. Drag the four data points so that the mean remains 5.00. Write down your new set of data points. Drag them again, while still keeping the mean at 5.00, and write down your new set of data points.

5. Press *Reset.* Add a new data point so that the mean is 6.00.

6. Drag one data point to 0. Drag your two other data points so that the mean remains 6.00. List all the ways you can find of doing this.

7. Place five data points anywhere on the number line, except at 4, so that they have a mean of 4.00.

8. Find two more sets of five data points that have a mean of 4.00 without any one of the data points having the value 4.

9. What strategy did you use to generate data sets with a mean of 4.00?

10. What can you say about data sets that have a mean of 4.00?

11. If you look at your answers to steps 7–10, you will see that you have created many different sets of data, all of which have a mean of 4.00. These may have included {1, 2, 3, 5, 9}, as well as {0, 3, 3, 6, 8} or {0, 0, 0, 10, 10}. Imagine that you added one more data point between 0 and 10 to each of these sets of data. Predict how much you think the mean might change.

12. Test your prediction from step 11. Describe your findings.

13. Drag the six data points to these locations: 2, 3, 4, 4, 5, and 6. Predict what will happen to the mean if one data point is added at 10. Test your prediction.

14. Can the mean increase to 10.00 by dragging in all the remaining data points from the left? Explain.

15. Suppose you find out that the mean value of the heights of the eight members of the family next door is 4 feet. Write down at least two different possibilities for the heights of each member of the family.

16. What kinds of things do you know for sure about the heights of each member of the family?

EXPLORE MORE

17. Press *Reset*. Imagine that your classmate placed a new data point far enough along the number line so that you couldn't see it on your screen. But you could see that the mean value of the three data points was 9.00. How could you figure out the value of the third data point?

INTRODUCE

Project the sketch for viewing by the class. Expect to spend about 5 minutes.

1. Open **Mean Meets the Median.gsp** and go to page "Median 5."
2. Explain, ***Today you're going to use Sketchpad to compare the behavior of the mean and the median. We'll start with a given data set and explore how changes to the values of that data set affect both the mean and the median. This will help us understand the difference between the two measures of central tendency and when it might be more appropriate to use one or the other. We'll look at questions such as "When is the mean greater than the median?" and "Can the mean and the median ever have the same value?" Before you begin, I'll demonstrate how to change a data set represented in the sketch.***
3. Point out that the vertical orange line represents the median value of the five data points. The line currently goes right through the data point at 4.0. Drag one of the data points along the axis until the median changes value. Drag a different point along the axis until the median changes again.
   - Ask students what they notice about the median value. They might describe it as always being in the middle of the data values, and they might also notice that the vertical line always passes through one of the data points. Make sure such intuitive ideas about the median are shared with the whole class in this introduction portion of the activity.
   - Draw attention to the fact that there are five data points. Later, students will investigate how to calculate the median of a data set that has an even number of data points.

DEVELOP

Expect students at computers to spend about 25 minutes.

4. Assign students to computers and tell them where to find **Mean Meets the Median.gsp.** Distribute the worksheet. Tell students to work through step 5, which wraps up the exploration of the median of five data points.
5. After most students have completed worksheet step 5, ask for examples of data sets with a median of 4.0. Illustrate them in Sketchpad. Ask students for different data sets (these might include data sets where there are two or more values on 4.0, as well as data sets that are either

clustered or spread out). The variety of data sets should show that even though the median value is always the middle value, it does not always look visually centered.

6. Ask students to work through step 13 and do the Explore More question if they have time. Let pairs work at their own pace. As you circulate, here are some things to notice.
   - Make sure students can articulate the difference between the way the median is calculated with five or six data points.
   - Encourage students to drag the different data points to different locations and not just focus on one data point.
   - For steps where students are asked to provide more than one example, encourage them to construct examples that are different. For example, the data sets {1, 2, 3, 4.1, 5}, {1, 2.1, 3, 4, 5}, and {1, 2, 3, 4, 5} are similar examples with medians of 3. However, {0, 0, 3, 3, 10} is quite different because it contains both repeated and extreme data points.
   - Encouraging students to see what their data sets have in common might help them construct different data sets. (Sometimes it's hard for them to think outside a particular strategy they have developed to generate new data sets.)
   - In worksheet step 13, you might ask students to try some sample data values so that they see that the waiting times for the patients who stayed all morning are much longer than for the other patients. Therefore the corresponding data points might be considered outliers.

SUMMARIZE

Project the sketch. Expect to spend about 15 minutes.

7. Gather the class. Students should have their worksheets with them. Begin the discussion by opening **Mean Meets the Median.gsp** and use it to support the class discussion.
8. Ask students to offer explanations for worksheet step 8. If students do not explicitly mention the words *odd* and *even*, probe their understanding by asking them what might happen in the case of data sets with 11 or 12 data points. They should be able to generalize that to find the median of an odd data set one simply locates the middle number, and to find the median of an even data set one must find the mean of the two middle numbers.

9. Now discuss the question in step 10, and ask students to give examples of three different ways to make the mean and median equal to each other. One might be to change the data point representing the median value, whereas another might be to increase or decrease one of the other values to push the mean closer to the median. Help students notice that there are many different data sets that will have the same mean and median, so that these two measures of central tendency do not completely describe a data set.

10. Students have now worked quite a bit on constructing different sets of data. ***Now we'll try to characterize the major differences between the mean and the median. Which measure of central tendency is more sensitive? Which might be more useful in different circumstances? Using the sketch, on page "Mean and Median" set the data values to 5.1, 5.4, 5.9, 5.9, and 6.2. Imagine these are the heights of five people lined up to go through security at the airport. The mean is lower than the median. Now suppose that one of the people lined up was a toddler, who measured 2.3. This new data value can be thought of as an outlier. How would the mean and median change?*** Drag one of the first values to 2.3. Now the mean will be much smaller than the median. This illustrates the way in which outliers can affect the mean without affecting the median at all.

11. Continue on to worksheet step 13. Ask for volunteers to describe their answers. Students should be able to recognize that the waiting times that are outliers influence the decision about hiring a new doctor, so the median might be a better measure. ***Are there other situations you can think of where you would want to consider the outliers?*** Encourage students to propose a variety of situations.

EXTEND

Suppose a teacher has just marked a mathematics test she gave to her students. Can you find a reason why she might prefer to know the mean of the test scores? Can you find a reason why she might want to know the median? Can you find a reason why she might want to know both? What other pieces of information might be useful to her? Encourage students to play with specific examples if they need to. If we want to know whether a test was too hard for most students, then the mean is helpful because it is sensitive to extreme values (in this case, to very low test scores).

However, the median might be a good measure if we don't care about extreme values (maybe some students were away, so we don't want their marks of 0 to affect the measure). It might also be helpful to know the range of the values. If the range is relatively wide, we might conclude that the test was not a very good one, because some students found it much too easy and others, much too hard.

You might also ask what data sets with equal means and medians have in common, and whether there are other measures of central tendency that have the advantages of each (for instance, dropping the highest and lowest values and finding the mean of the remaining values, as they do in the Olympics).

ANSWERS

1. Answers will vary. All data sets should include the value 4.0.
2. The median won't change as long as the data value remains greater than 4.0.
3. Once the data value becomes smaller than 4.0, the median will change because 4.0 is no longer the middle value.
4. Answers will vary. The data set should include two points with the same value as the median.
5. The median is the middle value when the data points are placed in order, but sometimes the middle value will not be visually centered among the data points.
6. Answers will vary.
7. Answers will vary. Some students might start by placing two data points on 7.5 and then two data points on each side of 7.5. Others might start with two data points whose average is 7.5 and then place two data points on each side of 7.5.
8. In step 1, with five data points the median is the middle value, but now the median is the average of the two middle values.

10. Answers will vary.
11. Answers will vary. Increasing the largest data value will not affect the median, but it will increase the mean.

12. Only the mean. Increasing the value of one data point does not change the middle value, but it does change the sum, and therefore the mean, of the values.

13. It's probably better to use the median because we don't want to include the extra waiting time of the patients who arrive early.

14. Answers will vary. The sum of the values of the two data points remains constant, so the mean is unaffected.

Mean Meets the Median

Name:

In this activity you'll investigate some properties of the median and compare its behavior to that of the mean.

EXPLORE

1. Open **Mean Meets the Median.gsp.** Go to page "Median 5." You should see data points at 1.0, 2.0, 4.0, 7.0 and 8.0, and a thick vertical orange line through 4.0 that represents the median value. Drag the data points. Write down different data sets that have a median of 4.0. What do all these data sets have in common besides having the same median?

2. Press *Reset.* Drag the data point at 8.0 to different locations. Describe all the values that data point can have without affecting the median value.

3. Press *Reset.* Drag the data point at 8.0 so that the median value changes. Why did it change?

4. Create a data set that has two data points on one side of the orange line and only one data point on the other side. Describe what's special about your data set.

5. Some people say that the median value is always in the middle. Why might this be a misleading way to characterize the median?

6. Go to page "Median 6." Drag the data points. Write down three data sets that all have a median of 4.0.

7. Describe a strategy for creating a data set that has a median of 7.5.

8. In step 1, you found that all your data sets had to contain the value 4.0. Is this still true? Explain.

9. Go to page "Mean and Median." Try to predict the value of the mean for the given data set. Then press *Show Mean.*

10. Find three different ways of changing the data points so that the mean and the median are equal to each other.

11. Find a data set in which the mean and median values are the same. Predict what will happen to each value if you increase your largest data value. Verify your prediction.

12. In many cases extremely small or large data values are called *outliers.* Will the mean or the median be more affected by outliers? Explain.

13. A doctor's office wanted to find out how long patients had to wait in order to see whether they needed to hire another doctor. After gathering their "waiting time" data, they learned that some patients had been dropped off first thing in the morning, even though their appointments weren't until late morning or early afternoon. Which measure of central tendency do you think would be more useful to the doctor's office?

EXPLORE MORE

14. Go to page "Median 6." Press *Show Mean.* Every time you move one of the data points, the mean changes. However, if you could move two data points at once, you could probably keep both the mean and the median the same. Press *Reset*. Select the data points 1.0 and 8.0 and choose **Edit | Action Buttons | Animation.** Set point *C* to animate in a forward direction and point *E* to animate in a backward direction. Press the button you just created to verify that your mean value stays the same. Explain why this works.

INTRODUCE

Project the sketch for viewing by the class. Expect to spend about 5 minutes.

1. Open **Quartile Craze.gsp** and go to page "9 Data Points." Enlarge the document window so it fills most of the screen.

2. Explain, ***Today you're going to use Sketchpad to explore a visual representation of data called the box-and-whisker plot, which provides an immediate visual sense of the center, the spread, and the overall range of distribution. We'll begin by looking at how the box and the whiskers are constructed. Then we'll vary our data set in order to gain a better understanding of questions such as "What do long whiskers tell us about our data?" and "Can the whiskers be uneven?" Before you begin, I'll demonstrate how to change a data set using the sketch.***

3. Go to page "Labeled Plot." Describe how the box-and-whisker plot is labeled for the given data set. It may be easiest to begin with the median, which divides the data into two sections and is a measure of central tendency that should be familiar to students. Press *Show Median.* Each segment is further divided into quartiles, with the lower quartile being the median of the data points to the left of the median, and the upper quartile being the median of the data points to the right of the median. Press *Show Quartiles.* These quartiles, along with the median, define the "box" part of the box-and-whisker plot. The length of the box is the interquartile range. Press *Show Box and IQR.* Finally, the whiskers extend from the quartiles toward the minimum data point (for the left whisker) and the maximum data point (for the right whisker). Press *Show Whiskers.* Tell the students, ***Refer back to this page as needed as you work on the worksheet.***

4. ***Now let's see how we can change the box-and-whisker plot by moving the data points.*** Show students how to drag individual data points.

   - Ask students what they notice about the values of the lower quartile, the median, and the upper quartile. Because there are only nine data points in this sketch, only the median will correspond to an actual data point.

   - Draw attention to the fact that the plot spans the entire data set, so extending a data point to be the largest or the smallest will change the length of the whiskers.

DEVELOP

Expect students at computers to spend about 20 minutes.

5. Assign students to computers and tell them where to locate **Quartile Craze.gsp.** Distribute the worksheet. Ask students to work through step 9 and do the Explore More if they have time. Here are some things to consider as you circulate.
   - Make sure students are dragging the data points and creating a variety of different box-and-whisker plots.
   - Make sure students understand how the two quartiles are calculated.
   - Help students describe their box-and-whisker plots qualitatively (comparing the size of the whiskers or the two sides of the box).
   - For worksheet step 9, encourage students to make good drawings of the box-and-whisker plots they create in Sketchpad.

SUMMARIZE

Project the sketch. Expect to spend about 5 minutes.

6. Gather the class. Students should have their worksheets with them. Begin the discussion by opening **Quartile Craze.gsp** and use it to support the class discussion.
7. Go to page "Sample Plots." Ask students to describe the important characteristics of each box-and-whisker plot and to provide an interpretation using an imagined set of data, such as test scores, heights, or another set of values. Make sure students compare the size of the whiskers as well as the shape of the box (including the position of the median in the box).
8. Ask a few students to re-create on the blackboard the box-and-whisker plots they obtained for worksheet step 9a. Compare these different plots and invite students to justify their solutions. There may be differences in interpretation, but the students should agree on the general shape of the plot. Do the same for worksheet steps 9b and 9c.
9. Tell students, ***This set of data had only 9 data points. How would increasing the data set to 10 or 11 points affect the construction of the box-and-whisker plot?*** It may be helpful to provide a sample set of data. For 11 data points (or any other odd number), the median will correspond to the middle value. With 10 data points (or any even number of data points), the median will correspond to the mean of the two middle values. You can extend this to analyze whether the

quartiles will correspond to data points. For both 10 and 11 data points they do, because in both cases the quartiles will correspond to the middle value of five data points. In general, the quartiles will correspond to data points for $4n + 2$ or for $4n + 3$ data points, but not for $4n$ or for $4n + 1$ data points.

EXTEND

The box-and-whisker plot describes the center and spread of a distribution. Another measure of spread uses the mean and standard deviation. It works better with distributions that are symmetric with no outliers, which is often the case with data describing natural phenomena. As you were working, you created many distributions that were not symmetric and that contained either one or two outliers. Create two different distributions in Sketchpad that are symmetric and that do not contain outliers. What can you say about the shape of the corresponding box-and-whisker plot? As students are working on this question, it may be helpful for them to refer back to the context of the test scores they explored in worksheet step 9. Encourage students once again to describe the box-and-whisker plots they create in terms of the relative size of the whiskers and the shape of the box. As a follow up, you might also discuss whether it's better to use the mean or the median to measure the center of a set of data.

ANSWERS

1. The median is the middle value of the data points.
2. The lower quartile is the median of the data points to the left of the median. The upper quartile is the median of the data points to the right of the median. Because there is an even number of data points on each side of the median, the quartiles are the mean of the two middle values.
4. The fifth largest data point directly affects the median. The data point with the minimum and maximum values directly affects the whiskers. The second and third largest data points contribute to the lower quartile's position, and the seventh and eighth largest data points contribute to the position of the upper quartile, which in turn also affect the lengths of the whiskers.

5. Yes, whiskers can be the same size. This will happen when the distance between the lower quartile and the minimum is the same as the distance between the upper quartile and the maximum.
6. The two sections of the box are unequal when the median is closer to one quartile than another.
7. The right whisker will be very long compared to the left one. The box will be small.
8. The whiskers will be about equal, as will the two sections of the box. The *IQR* will be greater than the whiskers.
9. Answers will vary.
   a. The plot should have long whiskers and the box should be small.
   b. The right whisker should be shorter than the left, and the right box should be smaller than the left.
   c. The two whiskers should be short. Either the right or the left box section should be much larger than the other.
10. Answers will vary. An outlier corresponds to a very long whisker.

Quartile Craze

Name:

In this activity you'll investigate some properties of box-and-whisker plots.

EXPLORE

1. Open **Quartile Craze.gsp** and go to page "Labeled Plot." Press *Show Median,* and then drag the data point at 13.2 past the data points on its left and right. How is the median determined?

2. Press *Show Quartiles,* and then drag the data points. How are the lower and upper quartiles determined? Explain why each quartile does not directly correspond to a data point.

3. Press the remaining two buttons and observe how the box and whiskers are formed.

4. Go to page "9 Data Points." Drag data points, one at a time. Which data points seem to affect the value of the median? Which data points seem to affect the lengths of the whiskers? Which data points seem to affect the position of the lower and upper quartiles?

5. Drag the data points again and focus on the whiskers. Can the whiskers ever be the same size? If so, find three different examples of when this happens. Explain what having equal whiskers means in terms of the quartiles and the maximum and minimum values.

6. Drag the data points again and focus on the box. You will notice that the two sections of the box aren't always the same size. What do you notice about your data set when those two sections of the box are not equal?

7. Predict what will happen if all the values except the maximum are clustered together. Describe what the resulting box-and-whisker plot would look like and verify your prediction.

8. Predict what would happen if all the values were more or less evenly spaced out. Describe what the resulting box-and-whisker plot would look like and verify your prediction.

9. Suppose the data set represents some test scores. For each situation presented here, create a corresponding box-and-whisker plot in Sketchpad and draw its basic shape.
   a. One student received a perfect score, another didn't answer any questions, and the rest received more or less the same mediocre score.

   b. Almost all the students received very high scores except for two, who only got half the questions right.

   c. About half the students did very well on the test and the rest did quite poorly.

EXPLORE MORE

10. You may have noticed that sometimes an orange dot appears above either the minimum or maximum value, or both. This occurs when those data points are *outliers,* data points that lie more than 1.5 times the IQR (interquartile range) from the edge of the box. Find examples of data sets having only one outlier and data sets having two. Describe what the box-and-whisker plot for each type of data set looks like.

3

Numbers, Expressions, and Equations in Grade 7

Open the Safe: Multiples and Factors

ACTIVITY NOTES

INTRODUCE

Project the sketch on a large-screen display for viewing by the class. Expect to spend about 30 minutes.

1. Open **Open the Safe Present.gsp** and go to page "The Safe." Explain, ***Today we're going to explore an unusual safe. It has a panel of 24 squares.***
   - Press the button on square 2. ***Which squares light up yellow?*** Some students may say that all even-numbered squares light up. Others may say that all multiples of 2 light up. Still other students may say that the second, fourth, and sixth columns light up.
   - Press the button on square 2 again. The class should note that all of the squares that lit up now turn off again, reverting to blue.
   - Press the button on square 4. Which squares light up? [All the multiples of 4]
   - Press the square 4 button again. Which squares turn off? [All those that were on]
2. Try another example or two. Students should be able to predict which squares will light up and to recognize them as multiples of the number on the button pressed.

   With several squares lit up, press the *Reset the Safe* button. Explain that this button resets all squares to their blue "off" position.
3. Now present a sequence in which two different numbers are pressed. Start by pressing *Reset the Safe*. Press the button on square 4. ***Now I'm going to press the button on square 6. What do you think the panel on the safe will look like?*** Have the class consider students' predictions. Press the button. Invite observations. Here are samples.

   All multiples of 6 changed their state from either off to on, or on to off.

   Squares 6 and 18 were off, so they turned on.

   Squares 12 and 24 were already on, so they turned back off.

1 Press	2 Press	3 Press	4 Press	5 Press	6 Press
7 Press	8 Press	9 Press	10 Press	11 Press	12 Press
13	14	15	16	17	18
19	20	21	22	23	24

1 Press	2 Press	3 Press	4 Press	5 Press	6 Press
7 Press	8 Press	9 Press	10 Press	11 Press	12 Press
13	14	15	16	17	18
19	20	21	22	23	24

4. To make sure that students understand how the safe works, try several more examples in which you start with all squares off and then press two different buttons in succession. Ask the class to predict the outcome before each button press.

5. Press *Reset the Safe.* Discuss the following two challenges.

   Which buttons can we press so that only squares 5 and 15 will be on? [5, and then 10]

   Which buttons can we press so that only squares 6 and 9 will be on? [6, then 9, then 12]

6. Prepare students to use the worksheet when they work at computers. Write the following on the board. ON: 12   OFF: 24

   Working at computers, you are going to solve problems in which certain squares must be on and others must be off in order to open a safe. Here is the combination for one safe: Square 12 must be on and square 24 must be off. Notice that this doesn't tell us anything about the other squares. Those squares can be either on or off. If you can figure out which buttons to press so that square 12 is on and square 24 is off, you'll be able to open the safe.

   Nothing actually opens on screen when the problem is solved. The context of the safe is for motivation.

   Solve this problem together, taking students' suggestions for buttons to press. One solution is to press button 12 followed by button 8. Be sure students observe that squares other than 12 are on when they are done.

   Ask students whether there are other solutions to the problem. There are several: 2-8, 4-8, and 6-8.

7. Distribute the worksheet. Explain that students should keep track of the buttons they press, as shown in the example. (Once students understand that they must press *Reset the Safe* before starting each new problem, they don't need to write *Reset.*) There are two spaces provided for each problem. If students find more than one solution for a problem, they should record two.

DEVELOP

Expect students at computers to spend about 40 minutes.

8. Assign students to computers and tell them where to access **Open the Safe.gsp.** Let students work at their own pace. Students should take turns being the button presser and the recorder.

9. As you circulate, observe students' approaches and the strategies they develop.

   - At first, students may not have any particular strategies for solving the problems. Allow them to experiment freely, and then begin to pose questions.

More examples of student reasoning are provided in the Summarize section that follows.

   - Students will likely think in terms of both multiples and factors. In problem 4, for example, some students may think about turning on square 14 by finding a number whose multiples include 14. This could be 1, 2, or 7. But 1, 2, and 7 can also be viewed as factors of 14. Thus some students may think about turning on square 14 by considering what its factors are.

   - If students don't like the result of a button press, remind them that pushing the same button again returns the panel to its previous state.

   - Notice the motivation level of students who still need to develop fast recall of the multiplication facts. Students realize that they have to know the facts to participate.

   - From time to time, ask, ***How many button presses did it take to solve this problem?*** Keep a record of the number of button presses students use where the class can see it. Some students may be intrigued and challenged to find a solution that matches or beats the number of presses used by others.

   - Problems 1–8 require no more than three button presses to solve. You can promote thinking about solution strategies by challenging students to try for no more than three presses. ***If you used more than three presses, reset the safe and try to open it with three or fewer presses.***

   - In worksheet step 9, students explain a strategy that they have used. This may be difficult for some students, but it's worth having them try to articulate their ideas.

Explore More

10. Worksheet step 11 is challenging. Students are asked to think of a problem that *cannot* be solved. As a hint, you might ask students to think about a problem in which square 1 is on. (For your information, when 1 is on, all of the prime-numbered squares that don't have a button—those greater than or equal to 13—are on too. The only way

to turn off one of these primes is to press 1 again. Thus there is no way for 1 to be on and a prime number greater than or equal to 13 to be off simultaneously.)

Some examples of problems that can't be solved are given in the Answers section.

11. Worksheet step 12 presents a set of challenging problems that can be solved. Students are likely to benefit most from tackling these problems at another time, after the Summarize discussion, but a few students may have time to solve one or two now. Explain that several problems require more than three button presses to solve.

SUMMARIZE

Project the sketch. Expect to spend about 20 minutes.

12. Gather the class. Students should have their worksheets. Open **Open the Safe Present.gsp.** Facilitate discussion of problems 1–8. Students should demonstrate by using the sketch and describe what each press accomplishes. Invite the class to help clarify descriptions. Also, encourage students to share multiple solutions to a problem. ***What is the same and what is different about these solutions?*** Some sample student descriptions are these.

    - Problem 2:

      Turning on 7 was easy—I just pressed 7. But pressing 7 also turned on 14, because 14 is a multiple of 7. To turn 14 off, I thought about what number had a multiple that was 14. The number 2 came to mind, and that worked great. Pressing 2 didn't turn off 7 because 7 isn't a multiple of 2.

    - Problem 3:

      Student A: *Turning on 9 was easy—I just pressed 9. But pressing 9 also turned on 18, because 18 is a multiple of 9. To turn 18 off, I had a choice of pressing either 2 or 6. I thought about pressing 3, but that wasn't a good idea. The number 9 is a multiple of 3, so pressing 3 would turn off 9 as well as 18.*

      Student B: *My answer is the same as yours, except I pressed 3 to turn on 9. When 9 turned on, so did 18. So, just like you, I needed to find a way to turn 18 off.*

- Problem 6:

  Student A: *Pressing 2 seemed like a good idea because it turned on both 2 and 24. I then pressed 4 to turn off 4. But because 24 is a multiple of 4, 24 turned off again. I needed to turn it back on without changing 2 and 4. Pressing 8 did the trick because 24 is a multiple of 8.*

  Student B: *I thought that the only way to turn on 2 was to press 2, but then I realized that I could also press 1. At this point, I was in exactly the same situation as you—4 was on and 24 was on. I had to press 4, just like you did, to turn 4 off.*

- Problem 7:

  There's only one way to turn on 17—press 1. Now I needed to turn 4 off by pressing 4. But pressing 4 also turned off 16. To turn 16 back on, I pressed 8.

13. Discuss worksheet step 9. Here are some sample student strategies.

    If a number needed to be on, I looked to see whether its square had a button. If it did, I pressed it to turn the square on.

    If a number greater than 12 needed to be on, I thought about what its factors were. I pressed one of its factors to turn the number on.

    Sometimes I needed to turn a number on or off without changing another number. I thought about a number that was a factor of the first number but not a factor of the second number.

    If a prime number greater than 12 needed to be on, I had to press 1.

14. If students have had time to tackle any of the Explore More challenges, have them share their discoveries now.

EXTEND

1. ***What other questions do you want to pose about the open-the-safe problem?*** Encourage student inquiry. The following mathematical questions are of interest.
   - Does the order of pushing buttons matter?
   - Why are the buttons only on the first half of the squares?
   - Would the safe be easier or harder to open if every square had a button?

- If you know which buttons to press to get one number on and another off, can you tell which buttons to push to get the second number on and the first number off?
- What would it be like to open a safe in which every square had a button, and pressing a button changed every number that was a *factor* of the button pressed?

2. Collect the problems students created in question 10 and share them with the class to solve. Students will enjoy trying to stump their friends!

ANSWERS

One possible solution is presented for each problem that follows. Other solutions are possible.

1. Press 12-8.
2. Press 7-2.
3. Press 9-6.
4. Press 7-3.
5. Press 12-9.
6. Press 2-4-6.
7. Press 1-4-8.
8. Press 1-2.

9–10. Answers will vary. See the Summarize section for possible responses.

11. The problem cannot be solved in the following cases:

    ON: 1 OFF: a prime number $\geq$ 13

    ON: a prime number $\geq$ 13 OFF: 1

    ON: a prime number $\geq$ 13 OFF: a different prime number $\geq$ 13

    ON: 8 OFF: 16

    ON: 16 OFF: 8

12. One possible solution is presented for each problem that follows. Other solutions are possible.

    a. Press 8-7-10.
    b. Press 5-10-4.
    c. Press 2-7-1.
    d. Press 7-3-9.
    e. Press 12-9-3-6.
    f. Press 9-8-4-12.
    g. Press 2-4-8-5-6.

Open the Safe

For GSP5 Name:

Open **Open the Safe.gsp.**

Go to page "The Safe."

Open the safe in each problem. Record the buttons you press.

Always start by pressing *Reset the Safe.*

If there is no information about a square, it can be either on or off.

If you find more than one solution, record two.

1. **ON:** 12 **OFF:** 24

   12 – 8 ______ 6 – 8 ______

2. **ON:** 7 **OFF:** 14

   ______ ______

3. **ON:** 9 **OFF:** 18

   ______ ______

4. **ON:** 14 **OFF:** 2, 21

   ______ ______

5. **ON:** 12, 18, 24 **OFF:** 6

   ______ ______

6. **ON:** 2, 24 **OFF:** 4

   ______ ______

7. **ON:** 16, 17 **OFF:** 4

   ______ ______

8. **ON:** all odd numbers **OFF:** all even numbers

_______________ _______________

9. Describe one strategy you used to solve the problems.

____

____

____

10. Write a problem of your own and its solution.
ON: _______ OFF: _______ Solution: _______

EXPLORE MORE

11. Now get tricky! Think of a problem that can't be solved. Put only one number in each blank space below. No matter what buttons are pressed, it should be impossible to open the safe.
ON: _______ **OFF:** _______

12. Try problems that are more challenging!
   a. **ON:** 8, 14, 20 **OFF:** 2

   _______________ _______________

   b. **ON:** 5, 15, 20 **OFF:** 10

   _______________ _______________

   c. **ON:** 14 **OFF:** 2, 7

   _______________ _______________

   d. **ON:** 7, 14 **OFF:** 9, 21

   _______________ _______________

   e. **ON:** 12, 18, 24 **OFF:** 6, 8, 9

   _______________ _______________

   f. **ON:** 18, 24 **OFF:** 6, 8, 12

   _______________ _______________

   g. **ON:** 2, 8, 20 **OFF:** 4, 24

   _______________ _______________

The Envelope: Adding and Subtracting Integers

ACTIVITY NOTES

INTRODUCE

Expect to spend very little time.

1. ***Today we're going to work with money. We'll pretend we're earning and spending money.*** As needed, review the meaning of the word *balance* for an amount of money and *debt* for money owed to someone else.

2. The worksheet and sketch are relatively self-explanatory. If you are concerned that your students may need more guidance, project the sketch for viewing by the class and spend a couple of minutes modeling how to change the value and how to press the buttons to add or subtract money.

DEVELOP

Expect students working at computers to spend about 20 minutes.

3. Assign students to computers and tell them where to locate **Envelope.gsp.** Distribute the worksheet. Tell students to work through step 24 and do the Explore More if they have time.

4. Let student pairs work at their own pace. As you circulate, be aware of students having difficulty. Ask questions that will help them move along.

   - In worksheet step 2, students may try to press the **+** or **–** button before entering the amount to be added or subtracted.
   - Students may resist entering numbers because they can do the calculations in their heads. Let them, but say, ***See whether you still feel that way when you get to the end of the worksheet.***
   - If students ask how you can have negative amounts of money in your envelope, you might say that you put an "I owe you" note in there for each debt.
   - If the point *Balance* moves off the part of the number line visible to students, suggest that they change the scale by dragging one of the scale numbers (other than 0).

5. You might ask selected students to write their responses to worksheet steps 12, 16, 20, and 24 on transparencies. They will share these strategies with the class later.

SUMMARIZE

Expect to spend about 10 minutes.

6. Reconvene the class, asking students to have their worksheets. Ask selected students to present their strategies from worksheet steps 12, 16, 20, and 24. Encourage honest but friendly critiques of the strategies.
7. Through the discussion, elicit these points.
   - Debt numbers are called *negative numbers;* the other nonzero numbers are *positive numbers.*
   - The word *negative* describes the number; the word *minus* means subtraction.
   - Negative numbers are on the "other side" of zero from positive numbers.
   - Together, the positive and negative whole numbers and zero are called the *integers.*
   - To add a positive number to any number, you move to the right or up on the number line.
   - To subtract a positive number from any number, you move to the left or down on the number line.
   - The directions in which you move are opposite if you're adding or subtracting a negative number.
   - In particular, adding a negative number is the same as subtracting the numerical value (ignoring the sign).
   - Subtracting a negative number is represented by moving to the right or up. It's the same as adding the numerical value (ignoring the sign). You may want to say that the numerical value is called the *absolute value* of the number. It's the (nonnegative) distance of the number from zero.
   - Emphasize that when students don't remember these rules, they can go back to the number line or to thinking about debts. With understanding, they can figure out appropriate rules for themselves.
8. ***What other questions can you ask about adding and subtracting integers in this way? You can ask a question even if you don't know the answer.*** Encourage all student inquiry. Here are some questions of mathematical interest.

- What if you added or subtracted amounts in between whole dollars?
- Can you have negative decimals or fractions?
- What if you're adding or subtracting more than two numbers?
- What situations other than debt can negative numbers represent?
- Are numbers with a positive sign really just the same as regular numbers?

ANSWERS

2. 19 dollars. $14 + 5 = 19$
3. 12 dollars. $19 - 7 = 12$
5. -3 dollars. $12 - 15 = -3$
6. The negative sign represents a debt, or the money you had to borrow.
7. The point moved down and ended up below zero on the number line.
8. 17 dollars. $-3 + 20 = 17$
9. -7 dollars. $17 + (-24) = -7$
10. The point moved down because you are spending money.
11. $17 - 24 = -7$
12. Answers will vary. Sample answer: Subtract the absolute values of the numbers and assign the sign of the number with the larger absolute value.
13. -11 dollars. $-7 + (-4) = -11$
14. The point moved down because you are going further into debt.
15. $-7 - 4 = 11$
16. Answers will vary. Sample answer: Add the absolute values of the numbers and assign a negative sign.
17. -6 dollars. $-11 - (-5) = -6$
18. The point moved up because you have less debt.

19. $-11 + 5 = -6$

20. Answers will vary. Sample answer: Change it to adding a positive number to a negative number.

21. 14 dollars. $-6 + 20 = 14$

22. 16 dollars. $14 - (-2) = 16$

23. $14 + 2 = 16$

24. Answers will vary. Sample answer: Yes, it's basically the same, except the first number is positive. Change it to adding a positive number to a positive number.

The Envelope

Name:

In this activity you'll develop strategies for adding and subtracting positive and negative numbers by observing how the amount of money changes in an envelope you keep hidden in your room.

EXPLORE

1. Open **Envelope.gsp** and go to page "Envelope." You should have $14 in your envelope.

2. You earn $5 by doing some yard work. Double-click the green value, type 5, and click OK. Then press + because you're putting money into your envelope.

   How much is in the envelope now? __________

   Write the calculation used to determine this amount.

3. You buy some food for $7. Change the green value to 7 and press – because you're removing money from your envelope.

   How much is in the envelope now? __________

   Write the calculation used to determine this amount.

4. Press *Show Number Line.* A point representing your balance—the amount of money in your envelope—appears on a vertical number line.

5. You want a new CD that costs $15. You don't have enough money in your envelope, so you borrow the rest from a friend. Change the green value to 15 and press –.

   What is the balance in the envelope now? __________

   Write the calculation used to determine this amount.

6. In your answer to step 5, what does the negative sign represent?

7. How did the point *Balance* move in step 5? Where did it end up?

8. To pay off your debt, you offer to clean a neighbor's house, and you are paid $20. Add this amount into the envelope. Sketchpad assumes that you pay off your debt first and put only what's left into the envelope.

   What is the balance in the envelope now? __________

   Write the calculation used to determine this amount.

9. You want to go to a concert that costs $24. Your aunt offers to loan you whatever you don't have. Instead of subtracting money from the envelope like you did in step 5, change the green value to −24 and press +.

   What is the balance in the envelope now? __________

   Write the calculation used to determine this amount.

10. How did the point *Balance* move in step 9? Why?

11. Use your observation from step 10 to write the calculation from step 9 another way.

12. In both steps 8 and 9, you added one positive and one negative number. Write a strategy for adding a positive number and a negative number.

13. At the concert, you want to buy a soda that costs $4. Your friend offers to lend you the money. To add more debt, add a negative number. Change the green value to −4 and press +.

    What is the balance in the envelope now? __________

    Write the calculation used to determine this amount.

14. How did the point *Balance* move in step 13? Why?

15. Use your observation from step 14 to write the calculation from step 13 another way.

16. Write a strategy for adding a negative number to a negative number.

17. Your aunt kindly cancels $5 of the debt you owe her. To remove debt, subtract a negative number. Change the green value to −5 and press −.
    What is the balance in the envelope now? __________
    Write the calculation used to determine this amount.

18. How did the point *Balance* move in step 17? Why?

19. Use your observation from step 18 to write the calculation from step 17 another way.

20. Write a strategy for subtracting a negative number from a negative number.

21. You get lucky and find a $20 bill on the street. Add it to your envelope.
    What is the balance in the envelope now? __________
    Write the calculation used to determine this amount.

22. In the last step, Sketchpad assumed that you first paid back both of your debts. While your friend accepted the $4 you owed for the soda, your aunt decided to cancel the $2 you still owed for the concert ticket. Remove this debt like you did in step 17.

    What is the balance in the envelope now? ___________

    Write the calculation used to determine this amount.

23. Write the calculation from step 22 another way.

24. Does your strategy from step 20 apply to subtracting a negative number from a positive number? If not, develop a strategy that works in both situations.

EXPLORE MORE

25. Go to page "Practice." Use your strategies to change the result on the right side of the equal sign. Sketchpad will let you know when you are correct. Keep pressing *New Expression* for more practice.

Right or Left: Adding and Subtracting Integers

ACTIVITY NOTES

INTRODUCE

Project the sketch for viewing by the class. Expect to spend about 5 minutes.

1. Open **Right or Left.gsp** and go to page "Addition." Enlarge the document window so it fills most of the screen.
2. Explain, ***Today you're going to use Sketchpad to add and subtract integers on a number line.*** You might remind students that integers are the set of whole numbers and their opposites, and ask them to give real-world examples of integers.
3. ***First I'll show you how the Sketchpad model works.*** Model worksheet steps 1–5. Here are some tips.
   - Introduce the addition problem in the sketch. ***What addition problem is shown?*** [8 + 5] ***Are the numbers in the problem positive or negative?*** [Both positive] ***Will the answer be to the right or to the left of the first number?*** [Right] ***Will it be to the right or left of zero on the number line?*** [Right] Ask a few volunteers to share their predictions. Then press *Add.* Press *Reset,* and then press *Add* a second time.

In this model each integer is represented by a vector. The term *arrow* is used because students may not be familiar with the term *vector* yet. A vector has both magnitude and direction, so practice with this model will help students understand how the signs of the addends come into play.

   - ***What does the first arrow represent?*** [8] ***What does the second arrow represent?*** [5] ***Why do both arrows point to the right?*** [Adding two positive numbers] ***How can you find the answer to 8 + 5 using the model?*** Students may make a response such as the following: *The second arrow starts from the end of the first arrow, so where the second arrow ends is the answer. The second arrow ends at 13.*
   - Now press *Reset,* and drag the circles to model 2 + (−6). ***By dragging the circles, I can change the problem. What is the problem now?*** [2 + (−6)] ***Why does the second arrow point left?*** [It represents a negative number.] ***Why does an arrow that represents a negative number point left?*** [Negative numbers are to the left of zero on a number line] ***How many units long is each arrow?*** [The arrow representing 2 is 2 units long; the arrow representing −6 is 6 units long.] ***Predict whether the answer will be to the right or to the left of zero.***
   - ***If I press* Show Steps, *we can see the solution modeled in steps.*** Press *Show Steps,* and then press each numbered button to see the model step by step. ***What is the answer to 2 + (−6)? How do you know using the model?*** Here is a sample student response: *The answer is −4*

because that is where the second arrow ended up after placing it at the end of the first arrow.

- You might consider doing another sample problem without pressing *Reset.* Students may gain more insight from leaving the arrows visible on the number line as they drag the circles to change the problem.

4. Ask students to consider the relationship between addition and subtraction of integers as they work. ***As you are adding and subtracting integers using the Sketchpad model, think about how adding and subtracting integers are related. How are they similar? How are they different?***

DEVELOP

Expect students at computers to spend about 35 minutes.

5. Assign students to computers and tell them where to locate **Right or Left.gsp.** Distribute the worksheet. Tell students to work through step 25 and do the Explore More if they have time. Encourage students to ask their neighbors for help if they are having difficulty with Sketchpad.

6. Let pairs work at their own pace. As you circulate, here are some things to notice.

   - In worksheet step 3 and for all activity questions, encourage students to write clear and detailed explanations using complete sentences. By clearly describing what they observe, students acquire a strong mental image of operations with integers. If time is limited, you might have students write their explanations for homework.

   - In worksheet step 8, have students predict what will happen in the Sketchpad model before pressing any buttons. ***What will the model of*** $-6 + (-3)$ ***look like? Why?*** Try to get students to concentrate on the behavior of the model rather than on the numeric answer.

   - In worksheet steps 10 and 11, encourage students to explore these questions by dragging to change the values of the integers without pressing *Reset.* Students can quickly view several problems before making a conjecture.

   - In worksheet step 12, students must interpret different parts of the Sketchpad model. As you walk around, observe students to be sure they understand each part and can model any problem they are given. When students successfully model all the problems, ask them to look for patterns. Students may notice that the sign of the answer is the same as the sign of the longer arrow. They may not recognize this as the integer with the greater absolute value; that's okay. Students are focusing on the visual model at this time. Discuss absolute value later.

- In worksheet step 14, students start subtracting integers. The concept of additive inverse is not explicitly named, but it plays a prominent role in the model. Ask students to think about why the second integer is flipped in a subtraction problem.
- As students are creating new subtraction problems, ask them to predict what the model will do each time before pressing the action buttons. ***Can you predict what the model of this subtraction problem will look like? Why do you think it will act that way?*** Students can test their conjectures using the step-by-step buttons.
- In worksheet step 24, ask students what patterns they see and how they could predict the answer from the two numbers being subtracted.
- In worksheet step 25, students are asked to write an addition problem using the same first number and the same answer. Students can test their addition problems by going to page "Addition." Switching back and forth between the two pages will reinforce the idea of using addition to rewrite a subtraction problem: To find the answer to a subtraction problem, you add the additive inverse (opposite) of the second number.
- If students have time for the Explore More, they will investigate the behavior of addition and subtraction independent of specific values, and they will use special cases to identify the position of zero on the number line.

SUMMARIZE

Project the sketch. Expect to spend about 20 minutes.

7. Gather the class. Students should have their worksheets with them. Begin the discussion by opening **Right or Left.gsp** and going to page "Addition." Work through the different types of addition problems with the class.
    - Have volunteers model the problems they recorded for worksheet step 9. ***What happens in the model when you add two negative integers?*** Students may make this sample response: *When adding two negative integers, the arrows both point left, so the answer is always*

negative. ***How does this compare to adding two positive integers?*** Students may reply that in both cases the arrows point in the same direction. With positive integers, the arrows point right. With negative integers, the arrows point left.

- Next have volunteers model the problems in worksheet step 12. ***What happens in the model when you add a positive and a negative integer?*** Students may make the following response: *If the negative integer is greater, the arrow pointing left will be longer, so the answer will be negative. If the positive integer is greater, the arrow pointing right will be longer, so the answer will be positive.*

- At this point, you may wish to introduce the term *absolute value* and the absolute value symbol. **Absolute value** ***is the distance a number is from zero. What represents the absolute value of a number in this model?*** Help students see that the length of an arrow is the distance from zero. ***What is the absolute value of −2?*** [2] ***What is the absolute value of 2?*** [2] Work through several problems with the class, each time focusing on the length of the arrow. Students should understand that opposites, or additive inverses, have the same absolute value. ***Can the absolute value of a number ever be negative?*** Students should realize that because distance is a positive value, the absolute value can never be negative.

- ***When adding a positive and a negative integer, how can you look at the numbers and tell whether the answer will be positive or negative?*** Students may make the following responses.

  The sign of the number with the longer arrow will be the sign of the answer.

  The sign of the number with the greater absolute value will be the sign of the sum.

8. Go to page "Subtraction." Have volunteers model subtracting two positive integers, a negative and a positive integer, a positive and a negative integer, and two negative integers. ***How are adding and subtracting integers related? How are they similar? How are they different?*** Students may respond with the following answers.

   When you subtract two integers, you flip the second number, so its arrow points the other way. You don't do that with addition.

In subtraction, after you flip the second number, the model is similar to addition. The answer is where the second arrow ends.

Subtraction is just adding the second number flipped.

In subtraction you are adding the opposite of the second number.

9. If time permits, discuss the Explore More. Have students explain how they determined the position of zero.

10. ***Explain the different ways you can get a negative answer when you subtract two integers.*** You may wish to have students respond individually in writing to this prompt. Here are the possible ways: If both integers are positive, the second integer must be greater than the first one. If the first integer is negative and the second integer is positive, the difference will be negative. If both integers are negative, the second integer's absolute value must be smaller than that of the first integer.

EXTEND

1. ***When you add two integers, does the order matter? In other words, is*** $-3 + 5$ ***the same as*** $5 + (-3)$***?*** Have students use the sketch to explain their answers. The order does not matter when you add two integers. The arrows determine how far you go and in which direction. It doesn't matter whether you follow the first arrow and then the second or whether you follow the second arrow and then the first.

2. ***When you subtract two integers, does the order matter? In other words, is*** $-3 - (-5)$ ***the same as*** $-5 - (-3)$***? Explain in terms of the model why your answer makes sense.*** The order does matter when you subtract integers, because only the second arrow is flipped. More sophisticated students will observe that the order matters only if the second integer is nonzero, because flipping zero has no effect.

3. ***What questions occurred to you while you were adding and subtracting integers?*** Encourage curiosity. Here are some sample student queries.

   Why do you flip the second arrow when subtracting?

   If you subtract a positive number from any number, is the result always to the left of the first number?

   Can this model work for multiplying and dividing integers?

ANSWERS

3. In their final positions, the second arrow starts from where the first arrow ends, and the answer (13) is at the end of the second arrow.

6. Answers will vary. Problems should include only positive integers.

7. Each bottom arrow is exactly the same size and direction as the corresponding top arrow.

8. The sum of $-6 + (-3)$ is -9.

9. Answers will vary. Problems should include only negative integers.

10. Whether adding two negative or two positive integers, both arrows go the same way, taking the sum farther away from the center of the number line (farther from zero). The difference is that the arrows go to the right when the numbers are positive and to the left when they are negative.

11. When you add two negative integers, you cannot get a positive sum. Both numbers take the sum in the negative direction from zero, so the sum must be negative.

12. $7 + (-4) = 3$ $\quad$ $-4 + 7 = 3$

    $-6 + 2 = -4$ $\quad$ $2 + (-6) = -4$

    $-3 + 7 = 4$ $\quad$ $3 + (-7) = -4$

    $2 + (-5) = -3$ $\quad$ $-2 + 5 = 3$

13. When you add a positive and a negative integer, the integer that has the greater absolute value tells you whether the answer will be positive or negative. In other words, the sign of the answer is the same as the sign of the longer arrow.

16. During the animation, the arrow for 5 flips from the right to the left. This shows which way the second arrow must go in order to subtract it from the first.

18. Answers will vary. This step creates the additive inverse by flipping the arrow.

19. Problems will vary. Problems should include those in which the second integer (*subtrahend*) is greater than the first integer (*minuend*).

20. If both integers are positive, the answer will be positive if the first integer is greater, and it will be negative if the second integer is greater.

21. Some students will write direct observations, and others will interpret those observations. Typical answers will be similar to these.

    Observation: In this problem, $4 - (-3)$, the second arrow starts out pointing to the left, so when it flips, it turns around and points to the right.

    Interpretation: The second number starts out negative, so when it flips, it becomes positive.

22. Problems will vary. The first integer is positive and the second integer is negative, so after flipping, both arrows point to the right. The answer must be positive.

23. Problems will vary. The first integer is negative and the second integer is positive, so after flipping, both arrows point to the left. The answer must be negative.

24. $7 - (-4) = 11$ $\quad -4 - 7 = -11$

    $-3 - 8 = -11$ $\quad -3 - (-8) = 5$

    $2 - (-7) = 9$ $\quad -2 - 7 = -9$

    $-6 - (-5) = -1$ $\quad -5 - (-6) = 1$

25. $7 + 4 = 11$ $\quad -4 + (-7) = -11$

    $-3 + (-8) = -11$ $\quad -3 + 8 = 5$

    $2 + 7 = 9$ $\quad -2 + (-7) = -9$

    $-6 + 5 = -1$ $\quad -5 + 6 = 1$

    Subtracting is the same as adding the opposite.

26. For addition, the sum moves in the same direction and at the same speed as the movement of either integer. For subtraction, the difference moves in the same direction and at the same speed as the movement of the minuend (a), but it moves in the opposite direction as the movement of the subtrahend (b).

27. For addition, move one integer until the sum is equal to the other integer. For subtraction, move the two integers to the same location so the difference is at zero, or move the subtrahend (b) until the difference equals the minuend (a).

Right or Left?

Name:

In this activity you'll add and subtract integers on a number line.

EXPLORE

1. Open **Right or Left.gsp** and go to page "Addition." If necessary, drag the circles to model the addition problem 8 + 5.

2. Press *Add.* Observe the model in action.

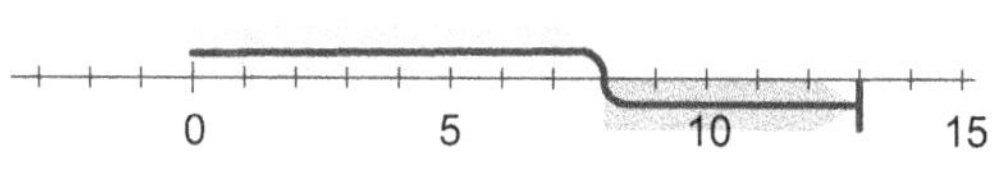

3. How does the final position of the arrows on the number line show the answer for this addition problem?

4. Press *Reset.* Drag the circles to model 2 + 6.

5. This time, press *Show Steps.* Then press each numbered button in order to see the model step by step.

6. Drag the circles to model another addition problem using only positive integers. Record your problem and its result.

7. How do the two top arrows in the sketch relate to the two bottom arrows?

8. Model −6 + (−3). What is the sum?

9. Model two more addition problems using negative integers. Record each problem and its result.

10. How is adding two negative integers similar to adding two positive integers? How is it different?

11. Can you add two negative integers and get a positive sum? Explain.

12. Model the following eight addition problems. Record each problem and its answer.

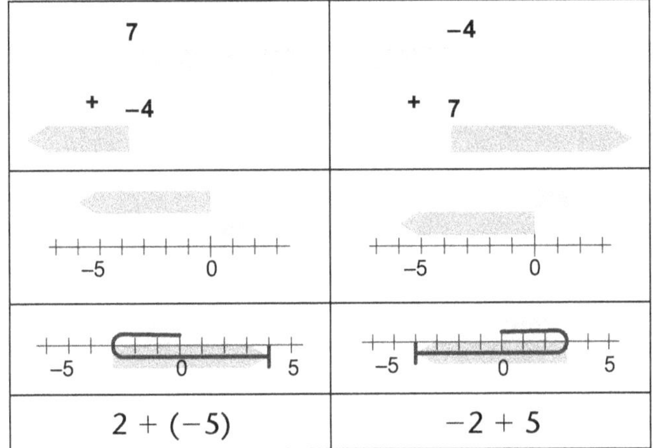

13. When you add a positive and a negative integer, how can you look at the numbers and tell whether the answer will be positive or negative?

Now you'll explore a subtraction model.

14. Go to page "Subtraction." If necessary, drag the circles to model the subtraction problem 8 − 5.

15. Press *Subtract.* Observe the model in action.

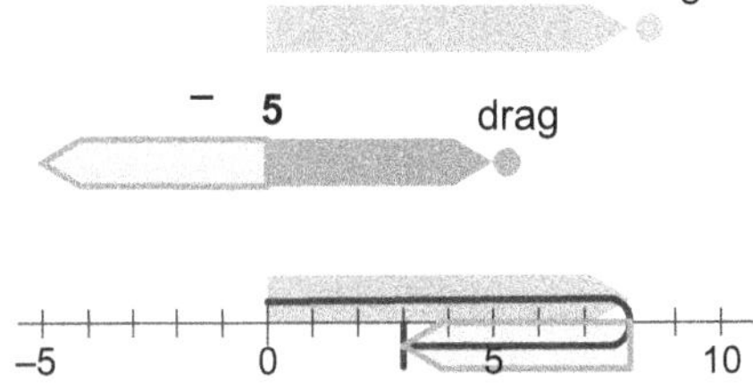

Right or Left?

continued

16. During the animation, what happens to the arrow for the integer 5? What does this show?

17. Press the *Reset* button. Then drag the circles to model 2 − 6.

18. This time, show the animation step by step. Describe in your own words what the step *3. Make Inverse* does.

19. Drag the circles to model two more subtraction problems that use positive integers but have a negative result. Record each problem and its result.

20. If both integers in a subtraction problem are positive, how can you tell whether the answer will be positive or negative?

21. Model 4 − (−3). What's different about the step *3. Make Inverse* this time?

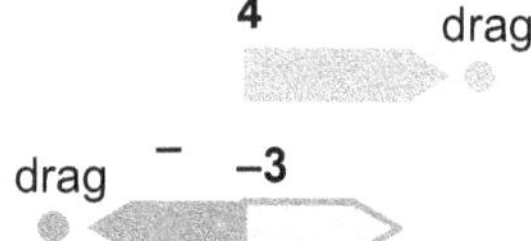

22. Model two more problems in which the first integer is positive and the second integer is negative. Record each problem. What do these models have in common?

23. Model three problems in which the first integer is negative and the second integer is positive. Record each problem. What do these models have in common?

24. Model the following eight subtraction problems. Record each problem and its answer.

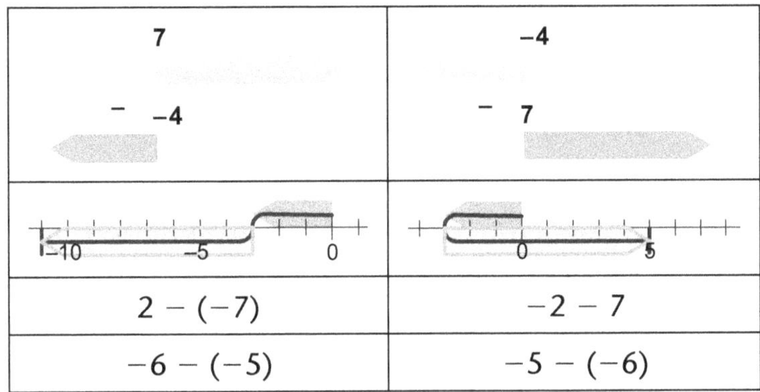

$2 - (-7)$	$-2 - 7$
$-6 - (-5)$	$-5 - (-6)$

25. For each subtraction problem in step 24, write an addition problem that has the same first integer and the same answer. What do you notice?

EXPLORE MORE

26. Go to page "Explore More." You will see two number lines, one that shows the sum $a + b$ and another that shows the difference $a - b$. With the numbers hidden, drag a and b and observe the behavior. How do addition and subtraction behave similarly? How do they behave differently?

27. On each number line, use your number sense to figure out where zero must be located. Drag a gold arrow to mark this location, and then press *Show Numbers* to check your answer.

Paper Cups: Connecting Slope and Unit Price

ACTIVITY NOTES

INTRODUCE

Project the sketch on a large-screen display. Expect this part of the activity to take about 10 minutes.

1. Start on page "Problem" of **Paper Cups.gsp.** Ask for a volunteer to read the situation description aloud.

   Strive for making sense. ***Can you describe the situation in your own words? What can you imagine about the characters and places and events involved?***

2. Go to page "Data." You might act surprised at the way the data are presented. ***Oh! The shopper recorded everything in a graph. Would you have done that?*** Students may have preferred to make lists or tables. ***Why do you think the shopper used a graph?*** Students might say, *To make a harder math problem.* Or, *It's easier to add new data—just make a point instead of writing down numbers.* Or, *It's easier to see patterns.*

3. ***Does the graph make sense to you? What would it take to make sense?*** Elicit the idea that students need to figure out what the coordinates mean and that labels on the axes would have helped. ***What might the coordinates of the points represent in the context of buying packages of paper cups?*** Students may be quick to decide that one coordinate is price, and consideration of the scales can convince them that the price coordinate must be the vertical coordinate. Again in context, the first coordinate probably represents the size of the package in some way. Students may propose to measure size in ounces. If other students don't challenge that idea, you might. ***How heavy is 30 ounces? Have you ever bought paper cups or plates?*** Some students may know from experience that packages of cups usually tell how many cups are included, and indeed that number would be important to someone buying cups for a party.

4. If your students are familiar with the ideas of dependent and independent variables, you might use those terms. If they aren't familiar with that terminology, but your curriculum calls for it, you might introduce the words now. The rest of this lesson does not refer to them, but you can insert them into the conversation frequently.

5. Questions may arise about the exact values of the coordinates. ***For now is it good enough to estimate their values?***

DEVELOP

Expect to spend about 10 minutes. Students may follow the class discussion at individual computers.

6. ***What problem are we trying to solve?*** Students may have to reread the description of the situation several times, because the instructions are buried. We need to figure out which packages of 9-ounce cups to buy.

7. ***What characteristics of the packages will you consider?*** Be open to a variety of ideas. Most students will want to think about price. Some may conjecture that the package with lowest price is represented by the point lowest in the coordinate plane—with the smallest second coordinate. If no one disagrees, try to include size in the discussion. ***I thought the cheapest was usually the largest.*** Bring the discussion to unit prices and their meaning. ***For these cups, what measurement unit would the unit price have?*** [Dollars per cup]

8. If time allows and students can turn to computers individually or in pairs, you might let them try to solve the problem on their own. No worksheet is needed. Otherwise, you can continue with the whole-class conversation. Either way, your goal is that students will tie the ratio of unit price to the slope of a line. ***Because a graph is geometric, is there some geometric meaning for a ratio like unit price? Is the cost over the number of cups the rise over the run of some line?*** It's a line from the origin to the point. And you want to find the point whose line has the smallest slope.

9. On the sketch, you or students can use the **Ray** tool to start a ray at the origin. Moving the pointer will swing the ray around to have various slopes. It's easy to see which data point the ray must go through to have the minimal slope.

10. Focus back on the problem. ***Have you solved the problem?*** Bring out the idea that you need to find the coordinates of the chosen point. Students may well see that the number of cups is 25 just by looking at the graph, but they will probably realize that they can't estimate the price very well. Choose **Measure | Coordinates** to display the coordinates of the chosen point.

11. If time allows, focus back on the unit price. ***So, what is the unit price?*** Students can divide 1.89 dollars by 25 cups to get 0.0756. ***What does that mean?*** [7.56 cents per cup]

SUMMARIZE

Expect to spend about 10 minutes.

12. If the students have been working individually or in pairs, reconvene the class for a summary. ***What have you learned?*** Record on display the points that students raise. They may include these.

    Graphs can help you understand data.

    It's difficult to understand graphs if the axes aren't labeled.

    The slope of the line from the origin to a point is the unit cost of the package represented by that point.

    The cheapest cups in unit price are in the package corresponding to the smallest slope.

    Students might also raise points about independent and dependent variables, if those topics arose in the introduction.

13. Introduce or remind students of the term *rate.* ***A ratio with* 1 *in the denominator is often called a* rate. *Do we have a rate here?*** [Yes, unit cost and slope are rates.] ***What other rates can you think of?*** Students will probably think of speed. Conversion factors such as feet per meter can be thought of as a special kind of rate.

14. ***What considerations other than least cost might you use to decide which package to buy?*** Students might suggest many ideas, including these.

    How many packages do I want to carry?

    What are the cups made of? [Some are paper and some are plastic.]

    By the time we get there, the store may be out of the cheapest cups, or it may have changed the prices.

15. ***What other questions can you ask?*** Brainstorming includes unproductive as well as productive questions. Here are some interesting examples.

    What's the most expensive package?

    Can you always solve rate problems with slopes of lines on a graph?

    Why is the slope the unit price?

    The goal is to pose problems, not solve them, but the first and third questions are good ones students can pursue in a deeper investigation.

EXTEND

Students can spend 5 to15 minutes or more working in pairs or individually.

You might offer the page "Practice" in **Paper Cups.gsp** to students who would benefit from more individual work on connecting rate ratios with slope ratios. The instructions accompany the sketch.

INTRODUCE

Project the sketch for viewing by the class. Expect to spend about 10 minutes.

1. Open **Signed Tiles.gsp** to page "Introduce." Enlarge the document window so it fills most of the screen.
2. Say, ***The tiles on this page represent an expression.*** Ask what the expression is. [$x + 2y + 4$] Manipulate the *x* slider and *y* slider so students see that the variables vary. ***What is the value of the expression when both x = 2 and y = 2?*** [10] ***When x = 1 and y = 3?*** [9]
3. ***We can check your answers by calculating the expression using Sketchpad's Calculator.*** Choose **Number | Calculate** and calculate the expression, clicking on the *x*- and *y*-values in the sketch to enter them in the expression. Once you have the expression, drag the sliders and say, ***To check the first answer, I drag the x slider to make x = 2, and I drag the y slider to make y = 2. We can see that the value of the expression for these x- and y-values is 10. Now if I change x to 1 and y to 3, the value of the expression is 9.***
4. Go to page "2x + 3y + 3" and page "New Expression" and say, ***As you work on this activity, the first couple of pages have tiles representing an expression you are to evaluate. Always evaluate the expression using paper and pencil first, then check your answer with Sketchpad's Calculator.***
5. Go to page "x^2 + 2x + 3" and say, ***Starting on page "x squared plus 2x plus 3," you will use custom tools to create your own expressions.*** Show the Custom Tools menu and the options available. Use the tools to create several example tiles, but not the actual expression on the page.

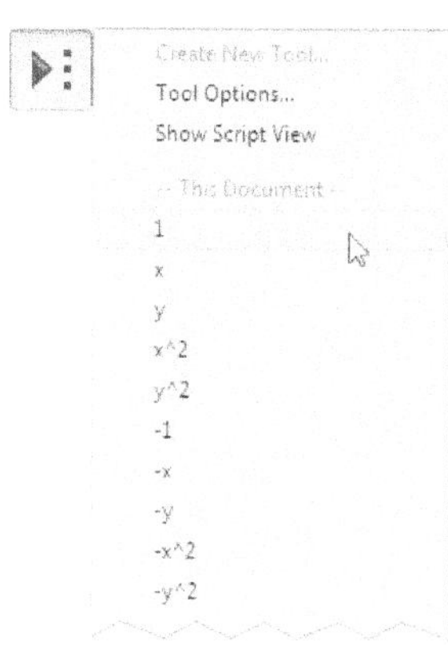

DEVELOP

Expect students at computers to spend about 25 minutes.

6. Assign student pairs to computers and tell them where to locate **Signed Tiles.gsp.** Distribute the worksheet and ask students to work on all steps. Encourage them to ask their classmates for help with Sketchpad.
7. Let pairs work at their own pace. As you circulate, here are some things to notice.
   - For students having trouble with the paper-and-pencil evaluations, suggest they write the expression replacing the variables with the

numbers before they multiply or combine terms. ***What is the value of this first term when x = 2?***

- On page "x^2 + 2x + 3" and beyond, students can choose to attach tiles to one another or not. If they make accidental tiles, remind them that they can choose **Edit | Undo**.

- On page "12," where $2x + y + 3$ is calculated, suggest that students leave x at one value and see what happens to the value of the expression as they move the y slider. ***What happens as y gets bigger? Why?*** Suggest they keep y constant and drag the x slider. ***How does dragging the x slider change the expression compared to dragging the y slider?***

- On page "13," where $3(x + 2) + y^2$ is calculated, students will have different ways of interpreting the parentheses and arranging the tiles. Make sure they have six units in their sketches. If they don't, ask them to explain how their sketch represents the expression and guide them with questions like ***Where are the three (x + 2)'s in your sketch?*** If their sketch is correct, you might ask, ***How does your arrangement represent the expression?***

- On page "14," where $4x - 2y + 4$ is calculated, make sure that students have used $-y$ tiles. Ask questions like ***What happens as y gets bigger? What happens as y gets smaller? Why? What does the −2y part do when y is negative?***

8. Encourage students who finish quickly to try the Explore More suggestion. They can challenge each other with expressions of their own devising.

SUMMARIZE

Expect this part of the activity to take about 10 minutes.

9. ***What have you learned from working with the tiles?*** Elicit ideas such as these.

   - The values of the variables determine the value of the expression. The value of an expression changes as the variables change.

   - The custom tools in the Sketchpad algebra tiles can create expressions in x, y, x^2, y^2 and 1 and negative versions of these.

   - The negative tile of a negative value represents a positive value.

- Expressions change more quickly when you drag the sliders when the slider variable is multiplied by a bigger number. (You could introduce the term *coefficient* here, if your curriculum uses the term.)

10. ***What other questions can you ask that might be explored?*** Encourage a variety such as these.

    Do some of these expressions have a smallest value or a largest value? Why?

    What kinds of expressions have a least value? What kinds have a largest value?

    Could we build an expression with x^3 or y^4 in it?

    Can expressions have more than two variables? Why are there only two variables in the custom tools?

    What are some of the expressions that can be made with the other custom tools?

ANSWERS

1. As the sliders are changed, tile lengths and numbers change.
2. $2x + 3y + 3 = 10$
3. Yes
4. a. 24 b. 26 c. 17
6. $3x + 6$
7. a. 18 b. 27 c. 3
10. a. 27 b. 66 c. 6
12. a. 14 b. 3 c. 9
13. a. 40 b. 16 c. 22
14. a. 6 b. -12 c. 24
15. Students' expressions will vary.

Signed Tiles

For GSP5 Name:

EXPLORE

1. Open the sketch **Signed Tiles.gsp.** The tiles on page "2x + 3y + 3" represent the expression $2x + 3y + 3$. Drag the x and y sliders. What happens to the tiles and numbers in the sketch?

2. Use paper and pencil to evaluate the expression $2x + 3y + 3$ for $x = 2$ and $y = 1$. What is the value?

3. To check your answer to question 2, choose **Number | Calculate.** Using ∗ for multiplication, enter the expression 2 ∗ x + 3 ∗ y + 3 in Sketchpad's Calculator. To enter the x and y in the expression, click on the x and the y in the sketch. Adjust the sliders to make $x = 3$ and $y = 1$. Does your answer to question 2 match Sketchpad's calculation?

4. Use paper and pencil to evaluate the expression $2x + 3y + 3$ for these values.

   a. $x = 3, y = 5$ b. $x = 1, y = 7$ c. $x = 4, y = 2$

5. Adjust the sliders to check your answers to step 4.

6. Go to page "New Expression." What expression do the tiles on this page represent?

7. Use paper and pencil to evaluate your expression in step 6 for

   a. $x = 4$ b. $x = 7$ c. $x = -1$

8. Choose **Number | Calculate** and calculate the expression 3 ∗ x + 6. Drag the x slider to check your answers to step 7.

9. On page "x^2 + 2x + 3," you'll use custom tools to create tiles. Press and hold the **Custom** tool icon to see the menu of tools. Choose a tool and click on the vertical line to add a tile. Use the tools to create the expression $x^2 + 2x + 3$.

10. Use paper and pencil to evaluate the expression $x^2 + 2x + 3$ for

    a. $x = 4$ b. $x = 7$ c. $x = -3$

11. Choose **Number | Calculate** and calculate the expression $x^2 + 2x + 3$. Drag the x slider to check your answers to step 10.

For questions 12–14, follow these steps.

Create tiles to represent the expression.

Use paper and pencil to evaluate the expression for each set of given values.

Use Sketchpad's Calculator to calculate the expression.

Drag sliders to check your answers to the question.

12. On page "12," create the expression $2x + y + 3$.

    a. $x = 3, y = 5$ b. $x = -2, y = 4$ c. $x = 4, y = -2$

13. On page "13," create the expression $3(x + 2) + y^2$.

    a. $x = 3, y = 5$ b. $x = -2, y = 4$ c. $x = 4, y = -2$

14. On page "14," create the expression $4x - 2y + 4$.

    a. $x = 3, y = 5$ b. $x = -2, y = 4$ c. $x = 4, y = -2$

EXPLORE MORE

15. On page "Explore More," use the tiles to create a complicated expression. Challenge yourself or a classmate to evaluate the expression for some given value or values. Check the answer using Sketchpad's Calculator.

Balancing: Solving Linear Equations

ACTIVITY NOTES

INTRODUCE

Project the sketch for viewing by the class. Expect to spend about 10 minutes.

1. Open **Balancing.gsp** and open page "Balance." Enlarge the document window so it fills most of the screen.
2. Explain, ***Today you're going to use a Sketchpad balance to model solving equations. Have you ever seen a balance like this one? How does it work?*** Let students share their understanding of balances. Some may have had experience in science classes using a triple-beam balance, which has a pan on one side and calibrated weights on the other, but few have probably used a balance with two pans. You might compare it to a seesaw, which will be familiar to most students. Make sure students understand that the heavier side will go down and that the two sides will balance only if they have the same weight.
3. Demonstrate how the model works, using only numbered weights. Remove a weight from the one side of the balance by dragging it into the storage bin. Then add some weights by dragging them from the storage bin. ***What do you see?*** [The balance tilts down toward the heavier side.]
4. Drag one x-weight to the left side and three 1-weights to the right side. They should balance (x is set equal to 3 on this page). ***What does this tell you about the weight of x?*** [It equals 3.] ***How would you write this as an equation?*** [$x = 3$] Use the **Text** tool to write the equation so that the equal sign is aligned with the center of the balance. Use a large font for visibility.

For a new text caption, use the **Text** tool. To move a caption, use the **Arrow** tool. To edit a caption, either click once with the **Text** tool or double-click with the **Arrow** tool.

5. Change back to the **Arrow** tool and add one 5-weight to the left side. ***What can I do to the right side to keep the balance?*** After students suggest adding 5, drag a 5-weight to the right side and write the new equation, $x + 5 = 8$. Use a new text caption so that you can align the equal signs.
6. Select the original equation and choose **Display | Hide Caption.** Then ask, ***What if we were trying to solve this equation? What could we do to isolate x, that is, to get x by itself?*** [Subtract 5 from both sides.] You want students to see that the process of solving an equation is the opposite of building one. Remove both 5-weights and unhide the original equation.
7. Add another x-weight on the left and ask, ***What can I do to the right side to keep the balance?*** Some students may suggest adding an x to the other side (adding the same thing to both sides) and others may suggest

adding three 1's to the other side (doubling both sides). Model both suggestions in turn, changing the caption to show the new equation ($2x = x + 3$, or $2x = 6$), and clarifying these two different interpretations. Show that to solve the equation $2x = 6$, students must reverse the process of doubling by finding half of what's on each side. ***What mathematical operation does this model?*** [Division by 2]

8. Go to page "A." ***Today you'll be solving equations like these. What is the original equation?*** [$x + 3 = 5$] Use the **Text** tool to write the equation in the sketch, and emphasize that students will do this on their worksheets. Then say, ***Your goal is to figure out how much x weighs. But you must do this while keeping the two sides balanced. What can you do to both sides of the balance to make it simpler?*** [Subtract three 1's from both sides.] Make sure students understand the idea of performing the same operation on both sides of the equation.

9. Say, ***The problems get more complex as you go, and you'll have to do operations besides subtraction. Some will require more than one step. Be sure to record the equations and the solution steps on your worksheets.*** If you wish to give more guidance, the page "Example" has an animation of how to solve the equation $3x + 2 = 8$ (press *Show Steps*). You might do the example together with the class, ask student pairs to review it on their own, or mention it to student pairs only if they need support.

DEVELOP

Expect to spend about 25 minutes.

10. Assign pairs to computers and tell them where to locate the sketch **Balancing.gsp.** Distribute the worksheet. Tell students to work through step 7 and do the Explore More if they have time.

11. Let pairs work at their own pace. As you circulate, here are some things to notice.

    - In worksheet step 3, equation B is designed to give students practice replacing a 5-weight with five 1-weights. Otherwise it is exactly like equation A.

    - In worksheet steps 4 and 5, check that students understand that equation C requires division whereas equations A and B involved subtraction.

- In worksheet steps 6 and 7, watch that students are systematically doing the same thing to both sides of the balance. Starting with equation D, the first two-step equation, some students might go directly to moving all but one x from one side and then manipulating the other side until they find the solution. This will work, but misses the point. Encourage students to articulate what they're doing in terms of mathematical operations (subtracting and dividing both sides of the equation by the same thing).
- Check that students are recording the equations and their solution steps on their worksheets.
- Some students are likely to work through the equations more quickly than others. Encourage them to experiment with the Explore More to build equations of their own.

SUMMARIZE

Expect to spend about 10 minutes.

12. Bring the class back together to discuss the strategies they discovered. ***What have you learned about solving equations?*** Bring out these objectives.
    - Equations might be solved by balancing, similar to working with a pan balance.
    - Subtracting the same number from both sides of an equation does not change the solution.
    - Dividing both sides of an equation by the same number does not change the solution. (You might remind students that they cannot divide by zero.)
13. ***What other questions can you ask that you may or may not be able to answer?*** Encourage all student curiosity. Mathematical questions of interest include these.
    - Can you also add and multiply both sides of an equation by the same thing?
    - Can you use balancing if the numbers aren't whole numbers?
    - Why do we need this balancing method if we can just guess and check solutions?
    - For what kinds of equations can we not guess and check solutions?

EXTEND

For practice with solving simple linear equations without the balance, randomly generated linear equations appear on page "Practice." Students who already have strong equation-solving skills should enjoy creating complex equations of their own on the page "Balance."

ANSWERS

1. $x + 3 = 5$
2. $x = 2$
3. $x + 3 = 7$
   $x = 4$
4. Even if you replace the 5-weight with five 1-weights, you must use division to find the solution.
   $5 = 2x$
   $2.5 = x$
5. Equation C modeled division. Equations A and B modeled subtraction.
6. Equation D: $2x + 1 = 5$
   $2x = 4$
   $x = 2$
7. Equation E: $2x + 8 = 3x + 5$
   $8 = x + 5$
   $x = 3$

   Equation F: $4x + 1 = x + 7$
   $3x + 1 = 7$
   $3x = 6$
   $x = 2$

   Equation G: $x + 7 = 4x + 4$
   $7 = 3x + 4$
   $3 = 3x$
   $x = 1$
8. Answers will vary.
9. Answers will vary.

Balancing

Name:

In this activity you'll use Sketchpad's balance model to solve equations.

EXPLORE

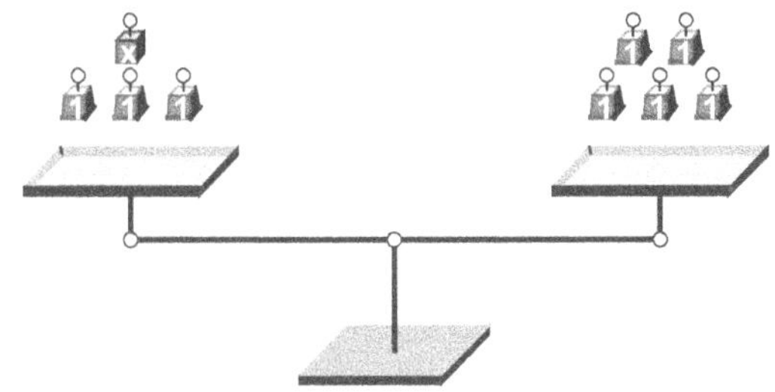

1. Open **Balancing.gsp.** Go to page "A." Write the equation represented by the balance.

2. To isolate the *x*, move 3 units off of each side of the balance. Does the scale still balance? Write the resulting solution.

3. Go to page "B." Write the equation. Drag items while keeping the two sides balanced to find the solution. Write the solution.

   Original equation:

   Solution:

4. Go to page "C." Write the equation. Explain why you can't drag items to find the solution. How can you find the solution for *x*?

5. What you did in equation C modeled a different operation than what you did in equations A and B. Which operations did you use to solve C? Which did you use to solve A and B?

6. Go to page "D." Solve this equation in two steps. After each step, the scale should balance. Record the balanced equations at each step here.

   Original equation:

   After first step:

   Solution:

7. Equations E, F, and G each require two or more steps. Use the balance to solve them and record your steps.

   Equation E:

   Equation F:

   Equation G:

EXPLORE MORE

8. Go to page "Balance." Build your own equation by putting *x*-weights and numbered weights on each side of the balance. Add 1-weights as necessary to balance the scale. Now solve your own equation, or challenge a classmate to solve it. When you're done, press *Change x-value* and repeat.

9. In the sketch, the *x*-value is always a whole number. Write an equation whose solution is a fraction. Show the solution steps.

10. Go to page "Practice" and try solving equations without using the model. Check you work by pressing *Show Solution;* then press *New Equation* for more practice.

Balancing with Balloons: Solving Equations with Negatives

ACTIVITY NOTES

INTRODUCE

Project the sketch for viewing by the class. Expect to spend about 10 minutes.

1. Students should have already completed the activity Balancing, which introduces the Sketchpad balance model using only positive weights. If not, ask, ***Have you ever seen a balance like this one? How does it work?*** Make sure students understand that the heavier side will go down and that the two sides will balance only if they have the same weight. You should also show examples involving only positive weights before introducing the balloons.

2. Open **Balancing with Balloons.gsp** and go to page "Balance." Enlarge the document window so it fills most of the screen. Explain, ***Today you're going to use a Sketchpad balance to model solving equations that include negative numbers and negative variables. Now there are balloons that pull up on the balance in addition to the weights that push down on the balance.*** Drag a 1-weight to one side of the balance, and then drag a -1-balloon to the same side. ***A negative one balloon pulls up just as much as a positive one-weight pushes down, so the scale remains balanced.*** Demonstrate that the same is true for an x-weight and a $-x$-balloon.

3. Drag one x-weight to the left side and four 1-weights to the right side. They should balance (x is set equal to 4 on this page). ***What does this tell you about the weight of x?*** [It equals 4.] ***How would you write this as an equation?*** [$x = 4$] Use the **Text** tool to write the equation so that the equal sign is aligned with the center of the balance. Use a large font for visibility.

For a new text caption, use the **Text** tool. To move a caption, use the **Arrow** tool. To edit a caption, either click once with the **Text** tool or double-click with the **Arrow** tool.

4. Add a -1-balloon on the left side and ask, ***What can I do to the right side to keep the balance?*** Students may suggest adding a -1-balloon to the right or removing a 1-weight from the right. Acknowledge both approaches, and then say, ***What happens if I do add a negative one balloon to the right?*** Besides balancing the scale, elicit the idea that a 1-weight and a -1-balloon cancel each other out. Drag the balloon so it is directly above the weight, select both of their points, and drag them simultaneously out to the storage bin.

5. Write the equation $x - 1 = 3$. Use a new text caption so that you can align the equal signs. Hide the original equation and ask, ***What if we were trying to solve this equation? What could we do to isolate x, that is, to get x by itself?*** [Add 1 to both sides.] Drag a 1-weight to both sides

of the equation. Then select both the weight and balloon on the left side, remove them, and unhide the original equation.

6. ***Today you'll be solving equations like these. What is the original equation? The problems get more complex as you go, and most will require more than one step. Be sure to record the equations and the solutions steps on your worksheets.*** If you wish to give more guidance, the page "Example" has an animation of how to solve the equation $-x + 4 = 2x - 2$ (press *Show Steps*). You might do the example together with the class, ask student pairs to review it on their own, or mention it only to student pairs if they need support.

DEVELOP

Expect to spend about 25 minutes.

7. Assign pairs to computers and tell them where to locate the sketch **Balancing with Balloons.gsp.** Distribute the worksheet. Tell students to work through step 6 and do the Explore More if they have time.

8. Let pairs work at their own pace. As you circulate, here are some things to notice.

    - Watch that students are systematically doing the same thing to both sides of the balance. Some students might go directly to moving all but one *x* from one side and then manipulating the other side until they find the solution. This will work, but misses the point. Encourage students to articulate what they're doing in terms of mathematical operations (subtracting and dividing both sides of the equation by the same thing).

    - In worksheet step 4, equation C requires students to replace a -5-balloon with five -1-balloons in order to divide it into two groups. Students may also just do the division mentally, which is fine.

    - In worksheet step 5, check that students understand that the solution to equation C is negative. The problem asks what this means in terms of the model. One way of thinking about it is that the *x*'s are balloons, which would mean that $-x$-balloons are weights. The main point of the question is to make students think, but don't focus too much on this limitation of the model.

    - In worksheet step 6, students will need to remove $-x$-balloons. Generally, students will find it easier to use positive *x*-weights to cancel out $-x$-balloons. If students don't, they will need to divide by a negative number on the last step, which is difficult to explain with this model. One way to think about it is that you isolate one $-x$-balloon, and so the solution for x must be the opposite of that value.

- Some students are likely to work through the equations more quickly than others. Encourage them to experiment with the Explore More to build equations of their own.

SUMMARIZE

Expect to spend about 10 minutes.

9. Bring the class back together to discuss the strategies they discovered. ***What have you learned about solving equations?*** Bring out these objectives.
   - Equations might be solved by balancing, similar to working with a pan balance.
   - Adding and subtracting the same number from both sides of an equation does not change the solution.
   - Dividing both sides of an equation by the same number does not change the solution. (You might remind students that they cannot divide by zero.)
10. ***What other questions can you ask that you may or may not be able to answer?*** Encourage all student curiosity. Mathematical questions of interest include these.
    - What does it mean to have a negative value for the solution?
    - Can you also multiply both sides of an equation by the same thing?
    - Can you use balancing if the numbers aren't integers?

EXTEND

For practice with solving simple linear equations without the balance, randomly generated linear equations appear on page "Practice." Students who already have strong equation-solving skills should enjoy creating complex equations of their own on page "Balance."

ANSWERS

1. $x - 5 = 2$
2. Add a 5-weight to both sides, and then remove the 5-weight and -5-balloon from the left side.
   $x = 7$
3. $3 = 2x - 1$
   $4 = 2x$
   $x = 2$
4. $2x + 1 = -5$
   $2x = -6$
   $x = -3$
5. The solution to equation C is negative. In terms of the model, x's are balloons (and $-x$-balloons are weights). Mostly, this illustrates a limitation of the model.
6. Equation D: $7 - x = 2x + 1$
   $7 = 3x + 1$
   $6 = 3x$
   $x = 2$

   Equation E: $x - 1 = -3x - 5$
   $4x - 1 = -5$
   $4x = -4$
   $x = -1$

   Equation F: $-3x + 2 = -x + 10$
   $2 = 2x + 10$
   $-8 = 2x$
   $x = -4$

   Equation G: $-x - 11 = -4x - 2$
   $3x - 11 = -2$
   $3x = 9$
   $x = 3$
7. Answers will vary.
8. Answers will vary.

Balancing with Balloons

Name:

In this activity you'll use Sketchpad's balance model to solve equations that include negative numbers and negative variables, represented by balloons.

EXPLORE

1. Open **Balancing with Balloons.gsp** and go to page "A." Write the equation represented by the balance.

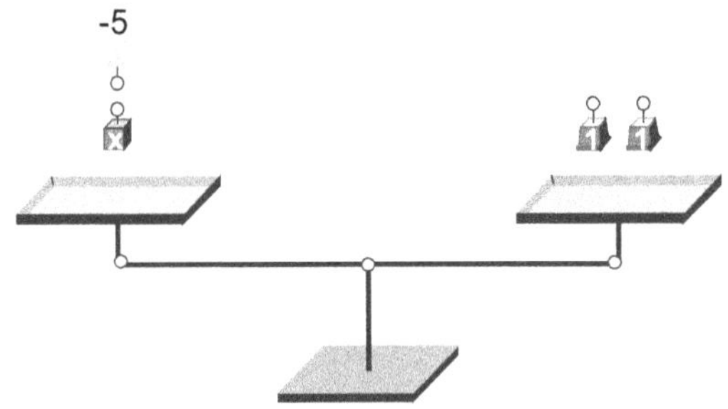

2. How can you isolate the *x*? Explain and write the resulting solution.

3. Go to page "B." Solve this equation in two steps. After each step, the scale should balance. Record the balanced equations at each step.

   Original equation:

   After first step:

   Solution:

4. Go to page "C." Solve this equation in two steps. After each step, the scale should balance. Record the balanced equations at each step.

   Original equation:

   After first step:

   Solution:

5. How is the solution to equation C different from the solutions to equations A and B? What does this mean in terms of the model?

Balancing with Balloons

continued

6. Equations D, E, F, and G each require two or more steps. Use the balance to solve them and record your steps.

   Equation D:

   Equation E:

   Equation F:

   Equation G:

EXPLORE MORE

7. Go to page "Balance." Build your own equation by putting weights and balloons on each side of the balance. Add 1-weights and –1-balloons as necessary to balance the scale. Now solve your own equation, or challenge a classmate to solve it. When you're done, press *Change x-value* and repeat.

8. In the sketch the *x*-value is always a whole number. Write an equation whose solution is a fraction. Show the solution steps.

9. Go to page "Practice" and try solving equations without using the model. Check your work by pressing *Show Solution,* then press *New Equation* for more practice.

Undoing: Solving Linear Equations

INTRODUCE

Project the sketch for viewing by the class. Expect to spend about 10 minutes.

1. Open **Undoing.gsp** and go to page "Explore." ***What calculations are being made in this diagram?*** [On the top branch, 3 is subtracted from 9, and the resulting 6 is multiplied by 2 to get 12. On the bottom branch, the 2 is subtracted from 12, and then 1 is subtracted from the result to get back to 9.]
2. Show how to drag one of the operations off the top branch and replace it with a different operation. ***What happened?*** [The bottom branch no longer undoes the top branch to get back to 9.]
3. Replace the original operation to return the diagram to the original form. Then demonstrate how to double-click on one of the numbers in the top branch and change its value. Again, the calculations on the bottom branch no longer result in the starting number.

DEVELOP

Expect students at computers to spend about 25 minutes.

4. Assign student pairs to computers and tell them where to locate **Undoing.gsp.** Distribute the worksheet. Ask students to work through step 8 and to go on to Explore More if they have time. Encourage students to ask each other for help with Sketchpad.
5. Let pairs work at their own pace. As you circulate, here are some things to notice.
   - In worksheet step 1, students should describe their reasoning, not simply state, "It doesn't work for all numbers."
   - If students pull the slider out very far to change the start number, they may have trouble getting it back onto the screen. Suggest they use **Edit | Undo.**
   - If students happen to choose **Display | Show All Hidden,** they should hide everything before they do anything else. If they have already clicked somewhere, they should undo back to the point where the controls were hidden.
   - In worksheet step 2, you might recommend that students begin with *Changed* and end with *Start.*

- You might suggest for worksheet step 3 that some students set up the top branch and challenge other students to find the corresponding bottom branch.

6. The Explore More activity extends undoing to equations whose coefficients or solutions are negative or fractions or decimals.

SUMMARIZE

Expect to spend about 10 minutes.

7. Reconvene the class. Students should have their worksheets with them.
8. ***What strategies did you find?*** Students likely will have found a successful strategy, but they may have difficulty articulating it. Use the terms *inverse operation* and *opposite order*. Reach class consensus on the main objective of the lesson: ***Some equations can be solved by undoing—considering the operations in opposite order and replacing each operation with its inverse operation.***
9. ***Does undoing equations remind you of anything in your everyday life?*** [Sample answer: Putting on and taking off shoes and socks—you put on socks before putting on shoes; to undo, you remove shoes before removing socks.]
10. ***What other questions might we ask?*** Encourage all student curiosity. Mathematical questions of interest students might raise include these.

    When page "Explore" was in its original format, why did Undone *keep changing by 2 when* Start *was changing by 1?*

    What's undoing good for?

    Are there equations that can't be solved this way?

ANSWERS

1. Answers will vary. The bottom branch does not undo the top branch for any starting number except 9. Each time you change *Start* by 1, *Undone* changes by 2.

2. $(Changed \div 2) + 3 = Start$

3. Answers will vary. The bottom equation should undo the top one.

4. Answers may vary. The operations will be considered in reverse order and replaced by their inverse operations.

5. Bottom: $(31 + 9) \div 5 = Start$, or $(31 + 9) \div 5 = 8$

   $Start = 8$

   Success of strategy and revised strategy may vary.

6–7. Answers will vary. The bottom equation should undo the top one, and the value of *Start* should check in the top equation.

8. a. $x = 1$

   b. $x = 6$

   c. $y = 17$

9. a. $x \approx 4.43$

   b. $x = -5$

   c. $n = \frac{13}{3}$

   d. $y = -8.48$

Undoing

Name:

In this activity you'll undo a linear equation to find an unknown value. You'll develop the strategy with known values before using it to find unknowns.

EXPLORE

1. Open **Undoing.gsp** and go to page "Explore." The calculations on the bottom branch of the diagram transform the 12 back into the start number, 9.

   The goal is not just to get back to 9. You want the bottom branch to undo the top branch for any start number. Drag the value of *Start*. Does the bottom branch undo the top branch for all start numbers? What patterns do you see?

2. On the bottom branch, drag the old operations out, drag new operations in, and change the numbers by double-clicking them until you get *Undone* to always match *Start* even when you drag the slider.

   The top branch can be represented by the equation

   $$(Start - 3) \times 2 = Changed$$

   What equation can represent your bottom branch?

3. Change the pink and blue operations and numbers on the top branch to set up a new situation. Then change the bottom branch until the undone number matches the start number for all start numbers.

   Top equation:

   Bottom equation:

4. What strategy do you think will always make the bottom branch undo the top?

5. To test your strategy, go to page "Solve." The start value is no longer visible. The equation for the top branch is $(Start \times 5) - 9 = 31$. Follow your strategy to drag in operations and change the numbers in the bottom branch to undo the top branch. Then press *Show Start* to see whether your result equals *Start.* Record your results.

   Bottom equation:

   Value of *Start*:

   Was your strategy successful? If not, what is your revised strategy?

6. Press *New Problem* and repeat step 5. Record your results. Use the letter x in place of the word *Start.*

   Top equation:

   Bottom equation:

   Value of x:

7. Press *New Problem* and repeat step 5. Record your results. Use the letter x in place of the word *Start.*

   Top equation:

   Bottom equation:

   Value of x:

8. Solve these equations. Use page "Explore" of the sketch if needed, but feel free to solve them without using the sketch. Write the equation that undoes the first one and calculate the value of the variable that represents the *Start* value.

   a. $(x + 6) \times 5 = 35$ b. $(x \div 3) + 7 = 9$ c. $(y - 5) \div 6 = 2$

EXPLORE MORE

9. The undoing process can solve equations even if the numbers are not positive integers. Solve these equations. Write each undoing equation and calculate the starting value, the value of the variable.

   a. $(x \times 3.5) + 2.1 = 17.6$ b. $(x + 7) \div 2 = 1$

   c. $\left(n - \frac{1}{3}\right) + \frac{7}{3} = \frac{19}{3}$ d. $\frac{y - 4.3}{7.1} = -1.8$

4

Geometry, Data, and Probability in Grade 7

Prism Dissection: Surface Area

INTRODUCE

Project the sketch for viewing by the class. Expect to spend about 5 minutes.

1. Show students the sample full-page net you cut out, and fold it to form a regular pentagonal prism. Explain, ***Today you will find the surface area of regular right prisms. You'll start with a pentagonal prism and then figure out a formula that works for other regular right prisms.***
2. Open **Prism Dissection.gsp.** Explain, ***On page "Prism," you'll review how to change the viewing angle and how to change the dimensions and number of sides.*** Model the use of the *spin, pitch,* and *roll* controls and how to change the number of sides.
3. ***What does surface area mean?*** Review the definition with the class. Here is a sample definition: The surface area of a prism is the combined area of all the faces of the prism. ***If you want to find the surface area of a prism, how can a net help you?*** Encourage students to explain why the area of the net is the same as the surface area of the prism itself.
4. If students have not previously used the Construct or Measure menus, briefly demonstrate how to construct a midpoint and how to measure the distance between two points. If students have not previously used the Calculator, show them how to choose **Number | Calculate** and how to click an object in the sketch to enter it into a calculation.

DEVELOP

Expect students at computers to spend about 30 minutes.

5. Assign students to computers and tell them where to locate **Prism Dissection.gsp.** Distribute the worksheet. Tell them, ***Once you've reviewed the controls on page "Prism," your next job will be to go to page "Base 1" and find the area of one base. If you can, try to figure out the measurements and calculations you need without using the hint.***
6. Give students time to work on their measurements and calculations in worksheet steps 5 and 6. If necessary, remind them to enter existing measurements into the Calculator by clicking on them rather than by typing the numbers in. Some students may be ready to go on to page "Base 2" before others have finished their calculations. If they do so, you can check their results for step 6 to make sure they answered 3.63. (All students should get this same result for a pentagonal base, no matter what size it is.)

7. When most students have finished worksheet step 6, call the class to attention and ask several students to report their values of *r*, the length of a side, and the step 6 calculation. Ask, ***What's interesting about these results from different-size bases?*** There's no need at this point for students to explain why the ratios are all the same, but it is important to bring this fact to their attention.

8. Tell students, ***If you used the number 5 in your calculation of the area, your calculation will work only for bases that are pentagons. On page "Base 2," do the same calculation, but in a way that will work for any polygon, no matter how many sides it has.***

9. Give students time to work on worksheet steps 7–11. Some students may be ready to go on to page "Faces" before others have finished their calculations. If they do so, you can check their results for step 10 to make sure they answered 3.14.

10. When most students have finished worksheet step 11, call the class to attention and ask several students to report their values of *r*, *n*, and the step 10 calculation. Ask, ***What's interesting about the calculated ratio now?*** Some students may mention that not only are the ratios very close to equal, but they get closer and closer to π as the number of sides gets larger and larger. Don't discuss this result in detail yet; give students time to think about it while they do the next few steps of the worksheet.

11. Tell students, ***Now that you've calculated the area of a base, you also need to calculate the area of a lateral face of the prism. Go on to page "Faces" and then to page "Area" when you're ready. If you finish early, try some of the Explore More questions.***

12. Give students time to work on worksheet steps 12–16. You might also consider assigning some or all of the Explore More, depending on your curriculum needs. If so, allot additional time.

SUMMARIZE

Project the sketch. Expect to spend about 10 minutes.

13. Have students discuss their results. Here are some questions for discussion.

    How did you find the area of the base?

    How did the net help you to understand the problem of finding the surface area?

You ended up with a formula that involved the values of n, r, h, and the length of a side of the polygon. Do you really need all of these values?

Encourage students to realize that these are interdependent. At the middle school level, you might discuss that for a given number of sides, the side length increases in proportion to the value of *r*. As you drag *R*, observe the side length and note that the triangles used to find the area of the base remain similar. At the high school level, you can use trigonometry to explicitly relate the side length to the values of *r* and *n*.

What interesting things happen as n increases? Students will observe that the shape becomes more and more like a cylinder. ***What happens to the area of the base?*** Some students will have noticed that the ratio in worksheet step 10 is 3.14, and that as the number of sides increases, the ratio approaches π. Some also may have drawn the conclusion that the area of the polygon approaches πr^2, the area of a circle. ***What happens to the area of the lateral faces?*** Some students will have noticed that the ratio in worksheet step 15 approaches 2π. Some may also have drawn the conclusion that the area of the lateral faces approaches $2\pi rh$, the circumference of the circle multiplied by the height. (Explore More worksheet step 18 explicitly asks students to use these facts to develop a formula for the surface area of a cylinder.)

ANSWERS

3. Answers will vary.

For the area calculations in steps 5 and 8, students can partition the base into 5 triangles with bases that are the side length (represented by *s* here) or 10 triangles with bases that are half the side length (represented by *t* here).

5. $A = 5 \cdot (s \cdot r)/2$ (alternatively, $A = 10 \cdot (t \cdot r)/2$)
6. For a five-sided polygon, the ratio is 3.63 and does not depend on the size.
8. $A = n \cdot (s \cdot r)/2$ (alternatively, $A = 2n \cdot (t \cdot r)/2$)
9. The value of the ratio depends on *n*. Any two students using the same value of *n* should have the same ratio.

10. For approximately $n = 100$, the ratio is 3.14. As n becomes large, the ratio appears to approach π. This makes sense because the more sides the polygon has, the more it looks like a circle, and the closer the measurement r is to the radius of the circle. Because the area of a circle is πr^2, the ratio of the area to r^2 is π.

11. It makes sense to use r because r measures the radius of the limiting circle. In fact, when n is small, r is the radius of the inscribed circle, though students may not make this observation specifically.

13. The area of the lateral faces is $n \cdot s \cdot h$.

14. When you divide by $r \cdot h$, the resulting value changes with n, but not with r or h.

15. When the value of n is large, the ratio becomes approximately 2π. One explanation is that the face area is equal to the perimeter $(n \cdot s)$ multiplied by h. As n increases, the perimeter approaches the circumference of a circle, so the face area approaches $2\pi r$ multiplied by h. When this value is divided by rh, the result is 2π.

16. The total surface area of the prism is the sum of the base areas and lateral face areas, so it can be expressed as $A = 2(n(s \cdot r)/2) + n(s \cdot h) = n \cdot s \cdot r + n \cdot s \cdot h$. The product $n \cdot s$ is the perimeter of the base (p), so the formula could be written as $n = p \cdot r + p \cdot h$.

18. When n becomes large, $n \cdot s$ approaches $2\pi r$, so the area approaches $2\pi r^2 + 2\pi rh$.

19. The volume must equal the area of the base multiplied by the height: $V = n \cdot s \cdot r \cdot h$, approaching a limit of $V = \pi r^2 h$.

Prism Dissection

Name:

In this activity you will create a regular prism with your choice of the height, the number of sides, and the size of the base. Then you'll calculate its surface area.

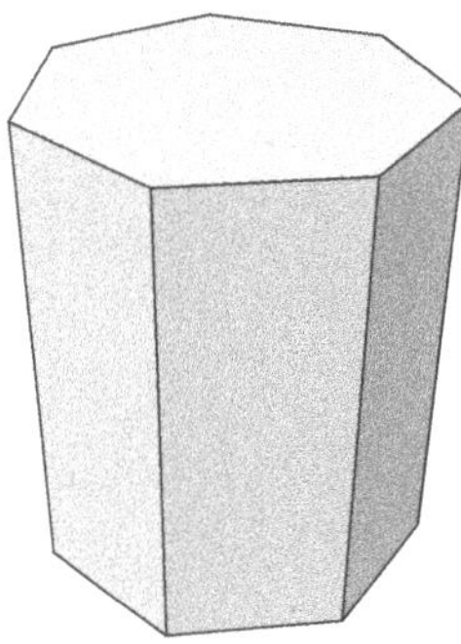

EXPLORE

1. Open **Prism Dissection.gsp** and go to page "Prism." Drag *spin, pitch,* and *roll* to view the regular prism from different angles.

2. Drag *N* to increase the number of sides. View the new prism from different angles. Drag *R* and *H* to change the shape of the prism.

3. How does the two-dimensional net correspond to the shape of the prism?

To find the surface area, you must find the area of both bases and of the lateral faces.

Base Area

4. On page "Base 1," the base has five sides. Change the size by dragging *R*. Construct the midpoint of the thick red side. Then measure the distance from *R* to the midpoint.

Double-click the distance measurement and label it *r*.

5. Find the area of the base. You'll divide the regular pentagon into simpler shapes, measure some distances, and do some calculations. (If you're stuck, pressing *Show Hint* can give you some ideas.) Write down your calculation and result.

6. Divide your resulting area by r^2. What value do you get? How does it compare with the results of other students?

7. On page "Base 2," measure distance r as you did in step 4. Change the number of sides and the size of the base.

8. Using n to represent the number of sides, write a calculation for the area of the base that will be correct for any value of n. Write down the result for your value of n.

9. Divide the area by r^2. What value do you get?

10. Increase the number of sides to more than 50. Now what is the value of the area divided by r^2? Have you seen this number before? Why do you think you get this value?

11. Why does it make sense to use the letter r to stand for the distance from the center to a side of the base?

Face Area

12. On page "Faces," drag *H, R,* and *N* to change the shape of the prism and the number of sides. Measure the distance r as you did in step 4. Then measure the height of the prism, and label it h.

13. Find the total area of the lateral faces. Write a calculation that will be correct for any value of *n*. Write down the result for your value of *n*.

14. Divide your resulting area by the product of *r* and *h*. What value do you get? Does this value change if you drag *R* or *H?* Does it change if you drag *N?*

15. What is this value when the number of sides is at least 50? Why?

Total Area

16. On page "Area," do the necessary constructions, measurements, and calculations to find the total area of the regular prism, including both bases and the lateral faces. Write down your calculation and result.

EXPLORE MORE

17. On page "Net," set *H, R,* and *N* to match the prism you made on page "Area." Choose **File | Page Setup,** set the page to print in landscape view, and then choose **File | Print Preview.** If necessary, change the scale so the net fits on one page. Then click **Print.** Cut out the net, fold along the lines, and glue or tape your three-dimensional prism together. Label each base and face with the area you calculated based on the measurements.

18. What does the prism look like when the number of sides is very large? How could you calculate the approximate surface area of this shape without using the value of *n?* (Your answers to steps 10 and 15 may be useful.)

19. Use your measurements to calculate the volume of the prism. Explain why you used the calculation you did. Then write your method as a formula.

20. On page "Explore More," you can experiment with the advanced controls that affect the look of this three-dimensional model.

INTRODUCE

Project the sketch for viewing by the class. Expect to spend about 5 minutes.

1. Students should have already completed the activities Prism Nets and Prism Dissection. Show students the sample full-page net you cut out, and fold it to form a regular pentagonal pyramid. Explain, ***Today you'll use Sketchpad to find the surface area of regular pyramids. You'll start with a pentagonal pyramid and then figure out a formula that works for other regular pyramids.***

2. Open **Pyramid Dissection.gsp.** Explain, ***On page "Pyramid," you'll review how to change the viewing angle and how to change the dimensions and number of sides.*** If students are not already familiar with the controls, model the use of *spin, pitch,* and *roll,* and the use of *N, R,* and *L.*

3. ***To find the surface area of a pyramid, what would you have to measure?*** Some students might provide a general description that you must measure the areas of the base and the five triangular lateral faces. Others might give more detail, describing how to measure the base and height of each triangle. ***How can a net help you?*** Students should see that the area of the net is equal to the surface area of the pyramid.

4. Holding up the folded pyramid, ask, ***Where would you measure the height of this pyramid?*** Encourage students to notice that there are two possible height measurements: the vertical distance from the center of the base to the vertex (height of the pyramid) and the distance from the base of a lateral face to the vertex (height of a triangle). Encourage students to discuss how these two different measurements might be useful, and discuss with them why it's important to avoid confusion by distinguishing the two heights. You may want to ask them to propose their own names for these measurements. Explain, ***On your worksheet, the distance from the base of a triangular face to the vertex is called the*** **slant height** ***and is labeled l.***

5. Students should have previous experience with the Construct and Measure menus. You may want to briefly review how to construct a midpoint, how to measure the distance between two points, how to change the label of a measurement, and how to click on an object in the sketch to enter it into the Calculator.

DEVELOP

Expect students at computers to spend about 30 minutes.

6. Assign students to computers and tell them where to locate **Pyramid Dissection.gsp.** Distribute the worksheet. Tell them, ***First review the controls on page "Pyramid" and answer the questions in steps 3 and 4. Then you'll go to page "Base" and find the area of the base. Try to figure out the measurements and calculations you need without using the hint.***

7. Give students time to work on their measurements and calculations. If necessary, remind them to enter existing measurements into the Calculator by clicking on them rather than by typing in the numbers. Some students may be ready to go on to page "Faces" before others have finished their calculations. If they do so, you can check their results for worksheet steps 6, 7, and 9. For step 9, make sure they answered approximately 6.29 for the ratio of the perimeter to r, and approximately 3.14 for the ratio of the area to r^2. Also make sure they've written an explanation for why they got these values.

   If students have different answers for worksheet step 9, check to make sure that they used the value of n and the actual measurements rather than typing in numbers, so that their results are correct no matter how they change the number of sides.

8. When most students have finished worksheet step 9, call the class together and ask several students to report their measurements and calculations of perimeters and areas from worksheet steps 6 and 7. Ask, ***What values did you get for the two calculations in step 9?*** Encourage students to explain why the two ratios come out to 2π and π. You may want to ask them whether the values are exactly 2π and π, and if not, why not? How could they measure the percentage by which the ratios differ from 2π and π?

9. Ask students, ***When you did your calculations, why was it important to click on the measurements in the sketch rather than just typing in the number?*** Encourage students to observe that they want their calculations to be correct even when they change the pyramid dimensions or the number of sides.

10. Tell students, ***Now you'll calculate the area of the lateral faces of the pyramid. Go on to page "Faces" and then to page "Area" when you're ready. If you finish early, try some of the Explore More problems.***

11. Give students time to work on worksheet steps 10–13. In step 12, some students may need a hint; you can suggest that they see what happens if they divide their numeric answer by *rl*.

    Consider assigning some or all of the Explore More questions, depending on your curriculum needs. If you do so, allot additional time.

SUMMARIZE

Project the sketch. Expect to spend about 10 minutes.

12. Have students discuss their results. Here are some questions for discussion.

    How did you find the area of the base?

    How did the net help you to understand the problem of finding the surface area?

    You ended up with a formula that involved the values of n, r, l, and the length of a side of the base. Do you really need all of these values?

    Encourage students to realize that these are interdependent. At the middle school level, you might discuss that for a given number of sides, the side length increases in proportion to the value of *r*. As you drag *R*, observe the side length and note that the triangles used to find the area of the base remain similar. At the high school level, you can use trigonometry to relate the side length explicitly to the values of *r* and *n*.

13. ***What interesting things happen as n increases?*** Students will observe that the shape becomes more and more like a cone. ***What happens to the area of the base?*** Students will have noticed that the ratio of area to r^2 in worksheet step 9 approaches π. They should be ready to draw the conclusion that the area of the polygon approaches πr^2, the area of a circle. ***What happens to the total area of the lateral faces?*** Some students will have noticed that the ratio of the perimeter to *r* in worksheet step 9 approaches 2π. Encourage students to explain how they could use the perimeter to make it easy to calculate the sum of the areas of these faces, by calculating *perimeter* × *slant height*/2. (Explore More worksheet step 15 explicitly asks students to use these facts to develop a formula for the surface area of a cone.)

EXTEND

What other questions might you ask about pyramids? Encourage all inquiry. Here are some ideas students might suggest.

What do pitch, roll, and spin really do? Are there other similar movements?

Why are the circumference and the area of a circle given by the familiar formulas? Is this activity a proof?

Can you do something like this to find the volume of the pyramid or cone?

Can you get the surface area of other curved figures, like a sphere, by taking some figure and increasing the number of faces?

ANSWERS

3. Student answers will vary. The *pitch* control allows you to see a top view, so that the pyramid looks like a regular polygon. In this view the *spin* control rotates the polygon about its center.
4. You cannot make the slant height l smaller than the value of r because the slant height must reach at least from the edge of the base to the center. When the slant height is equal to r, the pyramid is completely flat, with a height of 0. When the slant height is large compared to the radius, the pyramid is tall and skinny.
6. If students use s for the length of one side of the base, the perimeter is ns. Numeric results will vary and should change as students manipulate the dimensions.
7. One way to measure the area of the base is to think of it consisting of n triangles (as suggested by the hint), to measure the base (s) and height (r) of one triangle, and then use $A = sr/2$ to find its area. To find the area of the entire base, students can multiply by n, with the result that the total area is given by $nsr/2$.
9. The ratio of the perimeter to r is approximately 6.29—nearly 2π—and the ratio of the area to r^2 is approximately 3.14—nearly π. When $n = 60$, these values are accurate to about two decimal places.
11. The area of one lateral face is $sl/2$, and the sum of the areas of all these faces is $nsl/2$. If students use the perimeter p in their calculations, they will use the formula $pl/2$. Numeric answers will vary.

12. When the number of sides is large ($n \geq 50$), the perimeter approaches $2\pi r$, so the area of the sides approaches πrl.

13. The total area of the pyramid is the sum of the area of the base and the lateral faces: $nsr/2 + nsl/2$. Students may factor this to write it as $ns(r + l)/2$. Numeric results will vary.

15. When the value of n is large, the base becomes very nearly a circle, and the base area can be written as πr^2. The sum of the areas of the lateral faces approaches πrl, so the surface area of a cone is given by $\pi r^2 + \pi rl$, or $\pi r(r + l)$.

16. The height h of the pyramid (measured from the center of the base to the vertex) and the distance r form two legs of a right triangle, with the slant height l forming the hypotenuse. By the Pythagorean Theorem, $r^2 + h^2 = l^2$. If $l = 15$ cm and $r = 9$ cm, $h = 12$ cm. If $h = 5$ cm and $r = 12$ cm, $l = 13$ cm.

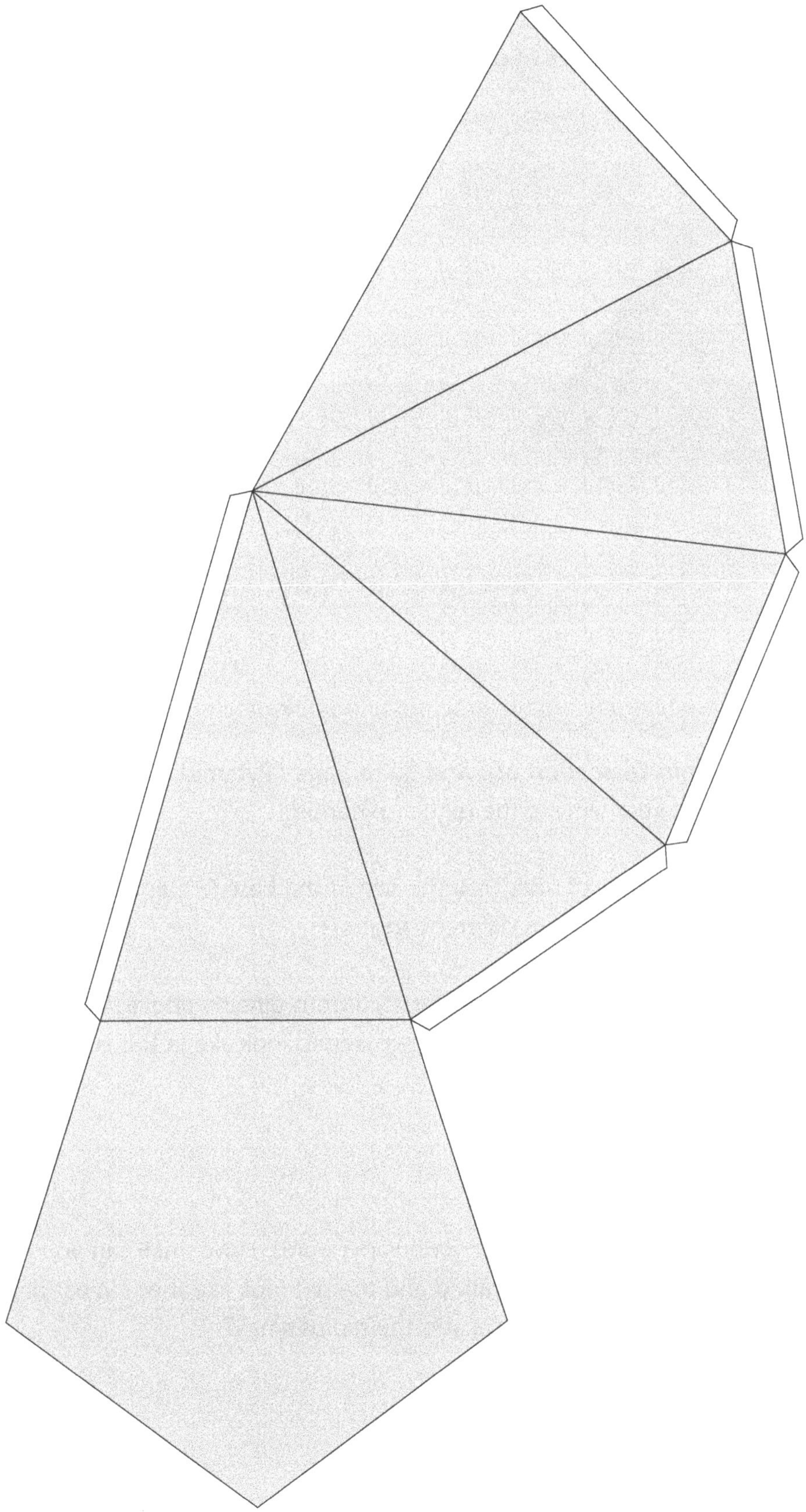

Pyramid Dissection

Name:

In this activity you'll create a regular pyramid with your choice of the height, the number of sides, and the size of the base. Then you'll calculate its surface area.

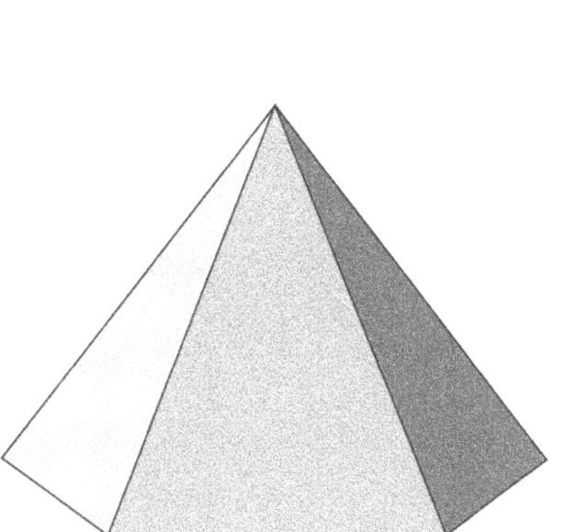

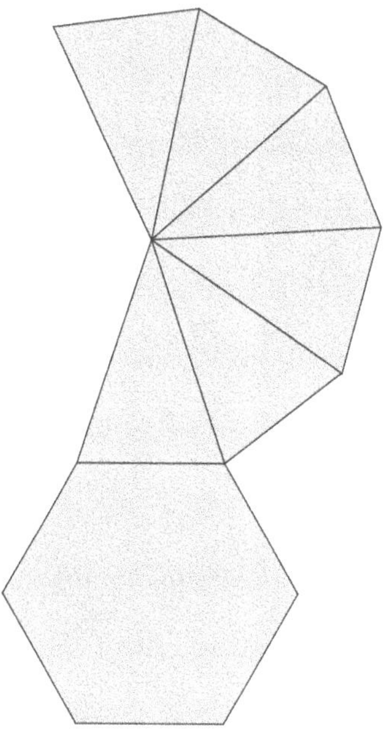

EXPLORE

1. Open **Pyramid Dissection.gsp** and go to page "Pyramid." Use *spin, pitch,* and *roll* to change your view of the regular pyramid.

2. Change the number of sides (*N*), the size of the base (*R*), and the slant height (*L*). View the pyramid from different angles.

3. Adjust the controls to look at the pyramid from directly above. Which control did you use to do this? What does the pyramid look like in this position? What does the *spin* control do now?

4. Explore the shape of the net for various pyramids. How small can you make the slant height? What do the pyramid and the net look like then? What happens if you make the slant height large and the radius small?

Pyramid Dissection

continued

To find the surface area, you must find the area of the base and of the lateral faces.

Base Area

5. Go to page "Base." Change the size of the base and change the number of sides. Measure the distance from *R* to the midpoint of the thick red side. Label the measurement *r*.

6. Measure one side of the base and have Sketchpad calculate the perimeter. Write down your calculation and result.

7. Find the area of the base. Imagine dividing the regular polygon into simpler shapes and do some measurements and calculations. (If you're stuck, press *Show Hint* to get some ideas.) Write down your calculation and result.

8. Change the number of sides and the size of the base, and make sure that your perimeter and area calculations seem reasonable. If not, fix them so that they work correctly for any base.

9. Increase the number of sides to more than 50. Divide the perimeter by r and divide the area by r^2. Have you seen these two numbers before? Why do you think you get these values?

Face Area

10. On page "Faces," drag *L*, *R*, and *N* to change the slant height, the size of the base, and the number of sides. Measure the distance r and side length s as you did in steps 5 and 6. Measure the slant height of the pyramid and label it l.

11. Find the total area of the lateral faces. Calculate a value that will be correct for any value of *n*. Write down the calculation and the result.

12. What is this value when the number of faces is at least 50? Why?

Total Area

13. On page "Area," do the necessary constructions, measurements, and calculations to find the total surface area of the regular pyramid, including the base and the lateral faces. Write down your calculation and result.

EXPLORE MORE

14. On page "Net," set *L, R,* and *N* to match the pyramid you made on page "Area." Choose **File | Print Preview** and make sure the net fits on one page. If necessary, click Scale To Fit Page, and then click Print. Label the base and each face with the area you calculated based on the measurements. Cut out the net, fold along the lines, and glue or tape your three-dimensional pyramid together.

15. What does the pyramid look like when the number of lateral faces is large? How could you calculate the surface area of this shape without using the value of *n*? (Your answers to steps 9 and 12 may be useful.)

16. The *height* (h) of a pyramid is defined as the vertical distance from the base to the *vertex.* In step 10, you measured the slant height l (the height of one of the lateral faces). How can you find the vertical height of the pyramid if you know l and r? (For instance, if $l = 15$ cm and $r = 9$ cm, what is h?) How could you find l if you know h and r? (For instance, if $h = 5$ cm and $r = 12$ cm, what is l?)

17. On page "Explore More," you can experiment with the advanced controls that affect the look of this three-dimensional model.

Dilation Designs: Proportions in Similar Polygons

ACTIVITY NOTES

INTRODUCE

Project the sketch for viewing by the class. Expect to spend about 5 minutes.

1. Open Sketchpad and enlarge the document window so it fills most of the screen. Explain, ***Today you're going to use Sketchpad to make a design that includes a polygon, and then you'll dilate it. By comparing the original polygon to the dilated image, you'll develop your own definition of similar polygons. First I'll show you how to measure a ratio using segments.*** As you demonstrate, make lines thick and labels large for visibility.
2. Construct two segments, select them, and choose **Measure | Ratio.** Drag one of the endpoints so that students see how the ratio changes. ***What does this number represent?*** [The ratio of the lengths of the two segments] Show students how the ratio is equal to 1 when the two segments are equal in length, and greater than or less than 1 when the segments are unequal in length. Point out that the order of selection is important. You might select the two segments in the opposite order and measure the ratio again to show this.
3. Say, ***Now I'll show you how to mark this ratio to dilate your design.*** Select the two segments and choose **Transform | Mark Segment Ratio.** Explain that when students are ready to dilate their design, this ratio will be used to determine the size of the dilation, but do not model the dilation itself.
4. If you want students to save their work, demonstrate choosing **File | Save As,** and let them know how to name and where to save their files.

DEVELOP

Expect this part of the activity to take about 15 minutes.

5. Assign students to computers. Distribute the worksheet. Tell students to work through step 15 and do the Explore More if they have time. Encourage students to ask their neighbors for help if they are having difficulty with the construction.
6. Let pairs work at their own pace. As you circulate, here are some things to notice.
   - In worksheet step 7, when students select **Dilate,** a pop-up window will prompt them to choose to dilate by fixed ratio or marked ratio. If they correctly marked the ratio of the segments, **Marked Ratio** will be selected.

- In worksheet step 12, students must select a side on the dilated image and then the corresponding side on the original polygon to create an equivalent ratio.
- In worksheet step 13, students are asked to drag points on the polygon, the ratio segments, and the center of dilation. By observing that the ratios remain equivalent, students should conclude that corresponding sides of similar polygons must have equivalent ratios.
- In worksheet step 14, students compare the angles of the polygons by dragging the center of dilation to each vertex. This should reinforce the idea that similar polygons have congruent angles.

7. If students will save their work, remind them where to save it now.

SUMMARIZE

Project the sketch. Expect to spend about 10 minutes.

8. Gather the class. Students should have their worksheets with them. Begin the discussion by opening **Dilation Designs Present.gsp** and using it to review worksheet steps 11–15.
9. Ask students to share their responses and discuss any other ideas or challenges that came up for them as they worked through the activity. Students will note that the figures are not generally the same size (so they're not congruent), but they are the same shape. ***What does it mean for polygons to be similar?*** [The ratios of corresponding sides will all be the same and corresponding angles will all be congruent.]
10. Decide as a class on a common definition of *similar polygons.* Sample definition: Two polygons are similar if all the corresponding angles are congruent and all the corresponding sides are proportional.
11. If time permits, discuss the Explore More. Discuss how the rays from the center of dilation through the vertices of the original image also pass through the vertices of the dilated image. This is a prerequisite for 1- and 2-point perspective drawing.
12. ***How do you know whether two shapes are similar?*** You may wish to have students respond individually in writing to this prompt.

EXTEND

1. Ask students to construct a pair of similar triangles without using a dilation. For example, construct a segment through two sides of a triangle that is parallel to the third side.
2. Ask students to construct two polygons that are not similar even though their corresponding angles are congruent. For example, the angles of any rectangle are congruent to corresponding angles of any square; but because the sides may not be proportional, the shapes are not necessarily similar.
3. Ask students to construct two polygons that are not similar even though their corresponding sides are proportional. For example, the sides of any rhombus are proportional to the corresponding sides of any square; but because the angles may not be congruent, the shapes are not necessarily similar.

ANSWERS

9. The polygons will coincide when the dilation scale factor is 1. This happens when the two segments defining the ratio have equal lengths.
13. The ratios of corresponding sides of similar polygons are equal.
14. The corresponding angles in similar polygons are congruent.
15. Similar polygons are polygons whose corresponding sides are proportional and whose corresponding angles are congruent.
16. Each ray through an original point passes through the corresponding image point.
17. The ratio of the distances from the center to an image point and from the center to the corresponding original point is the scale factor. For example, the image of a point dilated by a 2:1 scale factor is twice as far from the center of dilation as the original point.

Dilation Designs

Name:

In this activity you'll dilate a design and explore properties of similarity. You'll use what you discover to come up with a definition of similar polygons.

CONSTRUCT

1. In a new sketch, use segments to make a design that includes a polygon.

2. Construct a free point.

3. Label it *Center.*

4. Double-click the point to mark it as a center of dilation.

5. Construct two segments of different lengths.

6. Select the segments and choose **Transform | Mark Segment Ratio.** Remember the order you select them—it will determine the numerator and denominator of the ratio you'll find later in step 11.

7. Select your design and choose **Transform | Dilate.** Make sure **Marked Ratio** is selected in the dialog box. Then click **Dilate.**

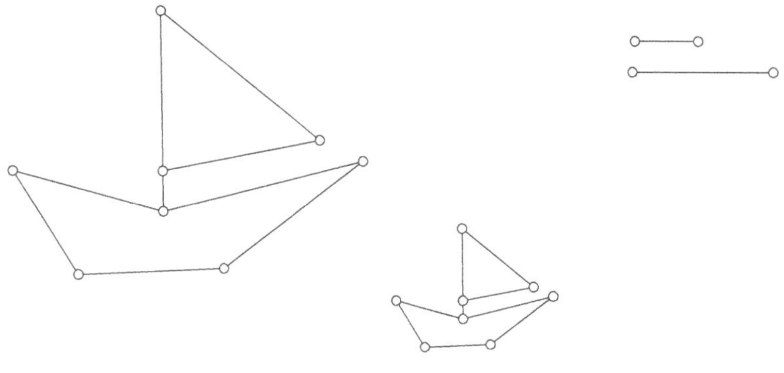

EXPLORE

8. Drag the center of dilation. Also, change the lengths of your two segments that define the ratio. Observe how these changes affect the similar designs.

9. How can you make the dilated image exactly the same size as the original design?

10. Drag the ratio segments so that the image is smaller than the original design.

11. Select the segments in order, as in step 6, and choose **Measure | Ratio.**

12. Measure the ratios of some corresponding sides of the polygons in your image and design.

13. Drag different points in the design, the ratio segment endpoints, and the center of dilation. What do you notice about the ratios of corresponding sides?

14. Drag the center of dilation onto each vertex of the polygon. What do you notice about the corresponding angles?

15. Use your observations to write a definition of similar polygons.

EXPLORE MORE

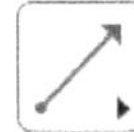

16. Construct rays from the center through each vertex of your original polygon. What other point does each ray pass through?

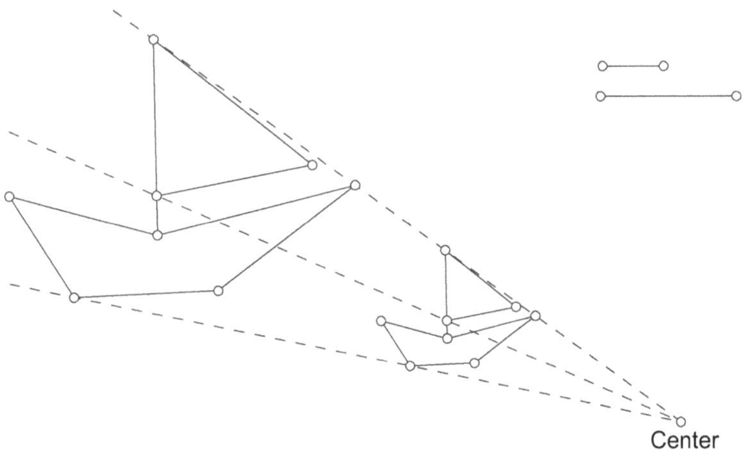

17. How does the distance between the center and the first polygon compare to the distance between the center and the second polygon?

Shape and Size: Exploring Similar Triangles

ACTIVITY NOTES

INTRODUCE

Project the sketch for viewing by the class. Expect to spend about 5 minutes.

1. Open **Shape and Size.gsp.** Go to page "Similar Triangles." Enlarge the document window so it fills most of the screen.
2. Explain, ***Today you're going to use Sketchpad to investigate the properties of similar triangles. What does it mean when two figures are similar?*** Take students' suggestions. If they only say that the two figures have the same shape, ask students whether the figures must be the same size. ***Do the two figures have to be the same size?*** Write an agreed-upon definition on the board: *Two figures are similar if they are identical in shape, but not necessarily the same size.* You might want to clarify that similarity requires more than just the same general shape: Not all triangles are similar. For your understanding, this activity does not consider reflections that are similar.
3. Direct students' attention to the sketch. Press *Show Vertices.* ***What are the names of these two triangles?*** [$\triangle ABC$ and $\triangle DEF$] If needed, review how to name triangles by their vertices. Drag vertices D and E on the blue triangle. ***What do you notice?*** Students may note that the position, orientation, and size of the blue triangle changes, but its shape and the sizes of its angles do not change. (The term *orientation* is used informally in this activity to refer to rotation, not the order of vertices.) Don't press students if they miss some observations; they will investigate the properties of similar triangles further during the activity.
4. ***In this activity you'll explore what makes these two triangles similar. You'll learn what the properties of similar triangles are and use them to find missing side lengths.***

DEVELOP

Expect students at computers to spend about 30 minutes.

5. Assign students to computers and tell them where to locate **Shape and Size.gsp.** Distribute the worksheet. Tell students to work through step 22 and do the Explore More if they have time. Encourage students to ask their neighbors for help if they are having difficulty with Sketchpad.
6. Let pairs work at their own pace. As you circulate, here are some things to notice.
   - In worksheet step 4, before students press *Blue to Yellow,* have them make the blue triangle dramatically smaller and in a different orientation than the yellow triangle. ***What happens during the***

animation? Help students identify the three stages of the animation: change in position (translation), change in orientation (rotation), and change in size (dilation). Students can also press *Yellow to Blue* to see the yellow triangle translate, rotate, and dilate.

- In worksheet step 6, as students are dragging the vertices of each triangle, ask them to try to predict which vertex of the blue triangle corresponds to each vertex of the yellow one. ***Can you tell which angle in the blue triangle corresponds to angle A in the yellow triangle? How about angle B? Angle C?*** Be sure students understand the term *corresponds* and define it as "matches," if necessary. Students can check their answers in the next step when they match up the triangles.
- In worksheet step 7, if students have trouble overlaying the triangles exactly, have them first press *Align Triangles* and then overlay the blue triangle on the yellow triangle.
- In worksheet step 9, review how to name angles by three points: a point on one side, the vertex point, and a point on the other side. Remind students that the vertex point must always be the second point named.
- In worksheet step 12, have students drag all the vertices. Encourage them to make triangles of different sizes. ***What happens to the angle measurements if you make the blue triangle really big? What happens to the angle measurements if you make the blue triangle really small? Try making the yellow triangle's angles all equal. What do you notice about the angle measures in the blue triangle?*** If necessary, remind students that the symbol $m\angle$ means "measure of the angle" and that angle measurements units are degrees.
- In worksheet step 14, students can choose the **Segment** tool and then choose **Edit | Select All Segments** to select all the segments in both triangles before measuring the lengths. After students measure the side lengths, ask them to drag the vertices and watch what happens to the measurements. ***If you drag the vertices of the blue triangle to change its side lengths, what happens to the side lengths in the yellow triangle?***
- In worksheet step 15, review the definition for *ratio* as needed: *A ratio is a comparison of two quantities by division.* Then check that students understand what the corresponding sides are in the two triangles.

Which side in the yellow triangle corresponds to side AB in the blue triangle? What other sides in the two triangles correspond?

- An alternate method for finding the ratio of side lengths is to select two sides (not their measure lengths) and choose **Measure | Ratio.** This would accomplish the same result as worksheet steps 14 and 15.
- In worksheet step 16, after students calculate all three ratios, have them drag the vertices of the triangles and observe what happens to the ratios. ***What do you notice about the ratios?*** Students should observe that the ratios change, but all three remain equal to each other.
- In worksheet step 18, remind students that a proportion is made up of two equal ratios. Tell students it's not fair to interchange the left and right sides of a proportion and count it as different.
- In worksheet step 21, students may find that their answers differ from other students' answers due to Sketchpad's rounding.
- If students have time for the Explore More, they will use similar triangles to find the length of a bridge. If students are stumped, help them identify the two similar triangles they can use. ***What triangle do you see in the scale drawing? How can you find the length of the two sides shown? What similar triangle can you use in the real world? What measurements do you know? What measurement are you trying to find out?***

SUMMARIZE

Project the sketch. Expect to spend about 10 minutes.

7. Gather the class. Students should have their worksheets with them. Begin the discussion by opening **Shape and Size.gsp** and going to page "Similar Triangles." ***What did you learn about the properties of similar triangles?*** Let volunteers come up to the computer and demonstrate. Students should respond that in similar triangles the corresponding angle measures and the ratios of corresponding side lengths are equal.
8. Then have students demonstrate how they solved the problems in worksheet steps 20 and 22. ***What proportion did you use to solve the problem? How did you know how to set up the proportion?*** Students should describe how they set up ratios of corresponding sides.

9. ***Suppose the ratio of corresponding sides of similar triangles is one to one. What do you know about the triangles?*** Students should recognize that the triangles are congruent. Demonstrate on the sketch for the class.

10. If time permits, discuss the Explore More. Ask students to describe the two similar triangles and what proportion they used to solve the problem. Triangle *ABC* in the scale drawing is similar to the triangle *ABC* in the real world. Students need to measure the lengths of the sides in the sketch to set up a proportion equivalent to $\frac{4.42}{x} = \frac{2.49}{500}$. Check that students understand that the answer is in meters.

11. ***Are all right triangles similar? Explain.*** You may wish to have students respond individually in writing to this prompt. Right triangles might have only one angle that is the same measure—the right angle. The two other angles measures may not be equal, so not all right triangles are similar.

EXTEND

What questions occurred to you about similar figures? Encourage curiosity. Here are some sample student queries.

What if you flipped one of the triangles over?

Is the ratio of two sides in one triangle the same as the ratio of the corresponding two sides in a similar triangle?

We just looked only at triangles. Do these properties hold true for other similar figures?

What other types of problems can you solve using similar triangles?

ANSWERS

3. As you drag points *D* and *E*, the angles of the blue triangle stay the same, but its size, position, and orientation change.

4. Both buttons show an animation that suggests the two triangles are the same shape.

7. The triangles match up when point *D* is on top of *A*, and *E* is on top of *B*.

9. When you make $\angle CAB$ in the yellow triangle very small, $\angle FDE$ in the blue triangle also becomes small.

10. When you drag point C far away from points A and B, point F moves far away from points D and E.
12. Angle DEF in the blue triangle corresponds to $\angle ABC$ in the yellow triangle, and $\angle EFD$ corresponds to $\angle BCA$. Angle CAB in the yellow triangle corresponds to $\angle FDE$ in the blue triangle.
13. The corresponding angles of similar triangles are equal in measure (or congruent).
17. The ratios of the three pairs of corresponding side lengths of similar triangles are equal.
18. Students should write these three proportions or their equivalents:

$$\frac{m\overline{AB}}{m\overline{DE}} = \frac{m\overline{CA}}{m\overline{FD}} \qquad \frac{m\overline{AB}}{m\overline{DE}} = \frac{m\overline{BC}}{m\overline{EF}} \qquad \frac{m\overline{BC}}{m\overline{EF}} = \frac{m\overline{CA}}{m\overline{FD}}$$

19. Corresponding angles of similar triangles are equal in measure (or congruent), and the ratios of corresponding side lengths are equal.
20. $m\overline{EF} = 6.7$ cm. This result is rounded off to the nearest tenth, as are all the distances in this activity.
21. Because of rounding, some students may end up with slightly different answers, such as $m\overline{EF} = 6.6$ cm.
22. $m\overline{AB} = 2.6$ cm. The proportion is $\frac{m\overline{AB}}{m\overline{DE}} = \frac{m\overline{CA}}{m\overline{FD}}$.
23. There are two similar triangles: One triangle appears in the scale drawing, with sides that students should measure in centimeters. The other straddles the banks of the river itself, with one side of 500 m and another side that is the unknown length of the bridge. One approach is to measure the distances in the scale drawing (using centimeters) and to set up a proportion involving the corresponding sides of the two triangles. The bridge must be 888 m in length.

Shape and Size

Name:

Two geometric figures are *similar* if they have the same shape, but not necessarily the same size. In this activity you'll investigate the properties of similar triangles and use proportions to find the missing sides of a pair of similar triangles.

EXPLORE

1. Open **Shape and Size.gsp** and go to page "Similar Triangles."

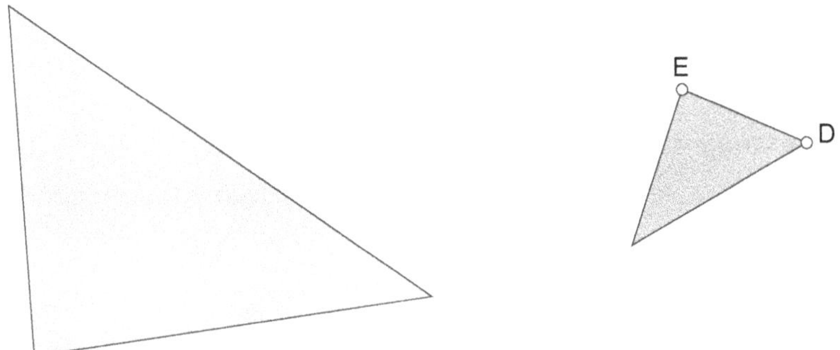

2. Drag points *D* and *E* to change the blue triangle.

3. As you drag the points, what changes and what stays the same?

4. Press *Blue to Yellow.* Then press *Yellow to Blue.* What does this seem to show about the two triangles?

5. To change the shape of the yellow triangle, first press *Show Vertices.*

6. Drag the vertices of each triangle. Observe what changes and what stays the same.

7. Drag point *D* and *E* until $\triangle DEF$ lies exactly on top of $\triangle ABC$. Where do you have to place points *D* and *E* to make the triangles match up?

8. Drag the triangles apart again.

Shape and Size

continued

9. Make $\angle CAB$ in the yellow triangle very small. What happens in the blue triangle?

10. Drag point *C* far away from points *A* and *B.* What happens in the blue triangle?

11. Press *Show Angle Measurements.*

12. Drag each vertex to see what happens to the angle measurements. Which angle in the blue triangle corresponds to $\angle ABC$ in the yellow triangle? Which angle corresponds to $\angle BCA$? Which angle in the yellow triangle corresponds to $\angle FDE$?

13. What can you conclude about the corresponding angles of similar triangles?

14. Select each side and choose **Measure | Length.**

15. Calculate the ratio of one pair of corresponding sides. To do so, choose **Number | Calculate** to open the Sketchpad Calculator. Click on a measurement to enter it into a calculation.

16. Calculate the ratios of the other two pairs of corresponding side lengths.

17. What can you conclude about the ratios of the corresponding side lengths of these triangles?

18. Complete the following proportion using the ratios you calculated in steps 15 and 16. Then write two other proportions using these ratios.

$$\frac{m\overline{AB}}{\underline{\quad\quad}} = \frac{\underline{\quad\quad}}{m\overline{DF}}$$

19. Using your answers to steps 13 and 17, write a summary of the properties of similar triangles.

20. If $m\overline{AB} = 3.0$ cm, $m\overline{BC} = 4.0$ cm, and $m\overline{DE} = 5.0$ cm, use the triangles to find $m\overline{EF}$. Set up a proportion to check your answer.

21. Compare you answer to step 20 with other students' answers. Then compare your sketches. Explain any similarities or differences.

22. If $m\overline{CA} = 4.5$ cm, $m\overline{DE} = 2.0$ cm, and $m\overline{FD} = 3.5$ cm, use the triangles to find $m\overline{AB}$. Set up a proportion to check your answer.

EXPLORE MORE

23. Go to page "Explore More." This sketch shows a scale drawing of a river and the location of a bridge that must be built between point *A* and point *B*. There is no easy way to directly measure the distance between the two points on opposite sides of the river. Use similar triangles and proportions to find the required length of the bridge.

Double Cross: Angles Formed by a Transversal

INTRODUCE

Project the sketch for viewing by the class. Expect to spend about 10 minutes.

1. Open Sketchpad and enlarge the document window so it fills most of the screen. Explain, ***Today you are going to use Sketchpad to explore the relationships among the angles formed when two parallel lines are intersected by a third line, called a* transversal. *I will demonstrate how to construct two parallel lines and a transversal and how to measure one angle formed by their intersection. Then you'll complete the same construction, but you'll measure all the angles formed and make conjectures based on your observations.***

2. As you demonstrate, make lines thick and labels large for visibility. First, model the construction in worksheet steps 1–8. Then model how to find an angle measure in worksheet step 9. Here are some tips.

Labels for points will automatically appear if you start with a new sketch and set **Edit | Preferences | Text** to **Show labels automatically: For all new points.**

To change a label, double-click the label. In the dialog box, type a new label and click **OK.**

   - Follow worksheet steps 1–4 to construct $\overleftrightarrow{AB}$ and a line parallel to $\overleftrightarrow{AB}$ through point *C*. Drag all three points to show that the lines always remain parallel.
   - In worksheet step 5, construct $\overleftrightarrow{CA}$. Drag points *C* and *A* to verify that the lines intersect at those points. Tell students to use **Edit | Undo** if they make a mistake in their constructions.
   - Construct points *D* through *H* and label them to match the figure in worksheet steps 7 and 8.
   - At this point, have students identify which lines are parallel and which line is the transversal. Define *transversal* as a line that intersects, or crosses, two or more lines. Students can name the lines several ways, depending on which points they use. ***Which lines are parallel?*** [$\overleftrightarrow{DE} \parallel \overleftrightarrow{GB}$ or $\overleftrightarrow{DC} \parallel \overleftrightarrow{GA}$ or $\overleftrightarrow{CE} \parallel \overleftrightarrow{AB}$] ***Which line is the transversal?*** [$\overleftrightarrow{FC}$, $\overleftrightarrow{FA}$, $\overleftrightarrow{FH}$, $\overleftrightarrow{CA}$, $\overleftrightarrow{CH}$, or $\overleftrightarrow{AH}$.]

You can set the angle units and precision in **Edit | Preferences | Units.**

   - Now model how to measure the size of an angle, such as $\angle CAB$. ***The points of the angle must be selected in the correct order, with the vertex as the second point.*** Identify the vertex as the point where the two sides of the angle meet. Select the three points, with point *A* second, and then choose **Measure | Angle.** Read the measure aloud. ***The* m *in front of the angle stands for "the measure of."***
   - Model how to deselect all objects by clicking in blank space in the sketch before selecting each new angle to measure. You can also deselect all objects by pressing the Esc key one or more times.

3. If you want students to save their work, demonstrate choosing **File | Save As,** and let them know how to name and where to save their sketches.

DEVELOP

Expect students at computers to spend about 20 minutes.

4. Assign students to computers and distribute the worksheet. Tell students to work through step 18 and do the Explore More if they have time. Encourage students to ask their neighbors for help if they are having difficulty with the construction.

5. Let pairs work at their own pace. As you circulate, here are some things to notice.

   - In worksheet steps 7 and 8, be sure students label their figures the same as the illustration on the worksheet. It will be important for later discussions that all student figures are labeled the same. Remind students that they can choose the **Text** tool and double-click the label to change it.
   - In worksheet steps 9, notice whether students correctly name the vertex as the second point when naming angles with three points. ***What is the vertex of the angle you are measuring? How do you know?*** If needed, tell them that the vertex is the point where the two sides of the angle meet.
   - If the **Measure | Angle** command is not available, have students check that only the three points that define the angle are selected.
   - Some students may have difficulty finding all eight angles because they name one angle more than once. Suggest that they drag each measurement to place it near the vertex of the angle just measured.
   - In worksheet step 11, observe which students have trouble identifying the angle types. Discuss the terms with these students. ***Which angles are in the interior, or between, the parallel lines? Which angles are in the exterior, or outside, the parallel lines? Which angles are on alternate, or opposite, sides of the transversal? Which angles are on the same side of the transversal? What does* corresponding *mean?*** [In the same relationship or similar position]
   - When students are stating the relationship between angle pairs, some students may have difficulty identifying the relationship between same-side interior angles and same-side exterior angles. Have students

choose **Number | Calculate** and click on one angle measurement, the + key, the other angle measurement, and **OK** to find the sum of the angle measures.

- In worksheet steps 13–18, be sure that students understand what they are investigating. ***You just explored the relationship between the angles formed by the intersection of two parallel lines and a transversal. Suppose you start with the relationship between the angles. What can you say about the lines?***
- In worksheet step 17, if the angle degrees are given in tenths or hundredths, students may have difficulty moving the lines until they have two sets of four congruent angles. Have students get the measurements as close as possible and then make their conjectures.
- If students have time for the Explore More, they will construct two congruent alternate interior angles by marking $\angle CAB$, selecting point C as the center for rotation, and rotating $\overleftrightarrow{AC}$ by the marked angle. The resulting new line is parallel to $\overleftrightarrow{AB}$.

6. If students will save their work, remind them where to save it now.

SUMMARIZE

Project the sketch. Expect to spend about 15 minutes.

7. Gather the class. Students should have their worksheets with them. Open **Double Cross Present.gsp** and go to page "Parallel Lines." Begin the discussion by talking about the construction of the parallel lines and the transversal in worksheet steps 1–5. ***You constructed two parallel lines and a transversal. Can you identify some parallel lines in our classroom? Can you find a transversal?*** Students may suggest the following ideas.

The top and bottom of the window form parallel lines. A window side is a transversal.

The left and right sides of the door are parallel. The top of the door is a transversal.

The top and bottom edges of my book are parallel. The side of my book forms a transversal.

The shelves on the bookshelf are parallel. The left and right sides of the bookshelf are transversals.

How do you know the lines are parallel? [They are the same distance apart; they don't intersect.] Write the notation for showing that two lines are parallel on chart paper: $\overleftrightarrow{AB} \parallel \overleftrightarrow{CD}$. ***How do you know which lines are transversals?*** [A transversal is a line that intersects two or more lines.] Write the definitions for *parallel lines* and *transversal* on the chart paper.

8. In worksheet steps 11 and 12, name an angle type—corresponding angles, alternate interior angles, alternate exterior angles, same-side interior angles, and same-side exterior angles—and have a volunteer come up and identify the angle pairs and state their relationship. When discussing same-side interior and same-side exterior angles, review the term *supplementary angles* and what it means.

9. If needed, move the transversal so that it is not perpendicular to the parallel lines. Then discuss other types of angles found in the figure. Ask students to find acute and obtuse angles. ***What do you notice about the relationship among these angles?*** [The acute angles all have the same measure; the obtuse angles all have the same measure; the acute and obtuse angles are supplementary.]

10. Go to page "Converse." Discuss worksheet steps 13–18. ***Did you find that the converse of your conjectures in step 11 were true? Explain.*** Explain what the term *converse* means, especially if your students are unfamiliar with this idea. Students should have discovered that if two lines are intersected by a transversal and one of these statements is true, then the lines are parallel.

    - Corresponding angles are congruent.
    - Alternate interior angles are congruent.
    - Alternate exterior angles are congruent.
    - Same-side interior angles are supplementary.
    - Same-side exterior angles are supplementary.

11. ***If one of the angles is a right angle, what can you say about the transversal? How do you know?*** [If one angle is right, then all angles are right angles, and the transversal is perpendicular to the parallel lines.]

12. You may wish to have students respond individually in writing to this prompt. ***Suppose a transversal intersects two parallel lines and you know the measure of one angle, do you know the measure of all angles formed? Explain.*** [Yes, three other angles are the same measure and the remaining four angles are supplementary to the angle.]

13. If time permits, discuss the Explore More. Ask students to share their findings.

ANSWERS

10. There are two sets of four congruent angles.

11. Students should complete the chart as shown.

Angle Type	Pair 1	Pair 2	Relationship
Corresponding	$\angle FCE$ and $\angle CAB$	See answer 12.	Congruent
Alternate interior	$\angle ECA$ and $\angle CAG$	$\angle DCA$ and $\angle CAB$	Congruent
Alternate exterior	$\angle FCE$ and $\angle HAG$	$\angle HAB$ and $\angle DCF$	Congruent
Same-side interior	$\angle ECA$ and $\angle BAC$	$\angle DCA$ and $\angle GAC$	Supplementary
Same-side exterior	$\angle FCD$ and $\angle HAG$	$\angle FCE$ and $\angle HAB$	Supplementary

12. There are four pairs of corresponding angles. Those not listed are $\angle ECA$ and $\angle BAH$, $\angle HAG$ and $\angle ACD$, and $\angle GAC$ and $\angle DCF$. Students should have recorded one of these pairs in the chart above.

18. The lines must be parallel.

24. Because the alternate interior angles were constructed to be congruent, the new line will be parallel to $\overleftrightarrow{AB}$.

Double Cross

 Name:

In this activity you'll explore the relationships among the angles formed when parallel lines are intersected by a third line, called a *transversal.*

CONSTRUCT

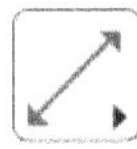

1. In a new sketch, construct $\overleftrightarrow{AB}$.

2. Construct point *C*, not on $\overleftrightarrow{AB}$.

3. Label the three points by clicking them in order from *A* through *C*.

4. Now you'll construct a line parallel to $\overleftrightarrow{AB}$ through point *C*. Select the line and the point and choose **Construct | Parallel Line.**

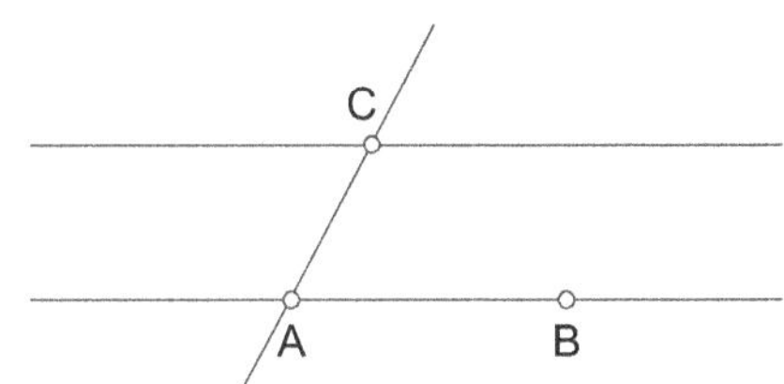

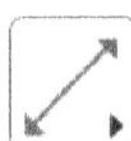

5. Construct $\overleftrightarrow{CA}$.

6. Drag points *C* and *A* to make sure the three lines are attached at those points.

7. Construct points *D, E, F, G,* and *H* as shown in the picture.

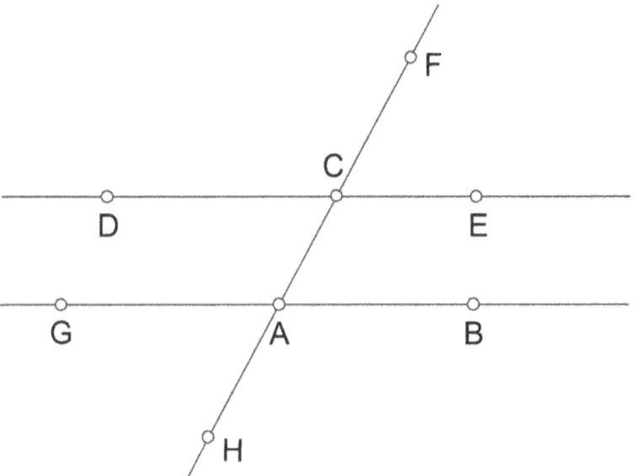

8. Label the points. You can double-click a label to change it.

EXPLORE

9. Now you'll measure the eight angles in your figure. Be systematic to make sure you don't measure the same angle twice. To measure an angle, select three points, with the vertex as your second point. Then choose **Measure | Angle.** Click in any blank space to deselect all objects after each measurement.

10. Drag point *A* or *B* and observe which angles stay congruent. Also drag the transversal $\overleftrightarrow{CA}$. (Be careful not to change the order of the points on your lines.) How many of the eight angles always appear to be congruent?

11. When two parallel lines are crossed by a transversal, the pairs of angles formed have specific names and properties. The chart shows one example of each type of angle pair. Fill in the chart with a second angle pair of each type, then explain how they are related.

Angle Type	Pair 1	Pair 2	Relationship
Corresponding	$\angle FCE$ and $\angle CAB$		
Alternate interior	$\angle ECA$ and $\angle CAG$		
Alternate exterior	$\angle FCE$ and $\angle HAG$		
Same-side interior	$\angle ECA$ and $\angle BAC$		
Same-side exterior	$\angle FCD$ and $\angle HAG$		

12. One of the angle types above has more than two pairs. Name that angle type and name the third and fourth pairs of angles of that type.

Angle Type	Pair 3	Pair 4	Relationship

13. Next, you'll investigate the converses of your conjectures. In a new sketch, draw two lines that are not quite parallel. Then construct a transversal.

14. Add points as needed.

15. Label the points *A* through *H* as shown in the picture.

16. Measure all eight angles formed by the three lines.

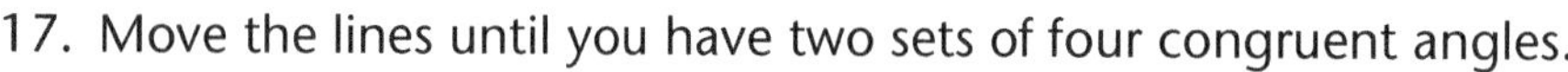

17. Move the lines until you have two sets of four congruent angles.

18. If two lines are crossed by a transversal so that corresponding angles, alternate interior angles, and alternate exterior angles are congruent, what can you say about the lines?

Double Cross

continued

EXPLORE MORE

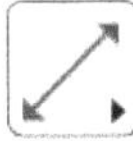

19. You can use the converse of the parallel lines conjecture to construct parallel lines. Construct a pair of intersecting lines $\overleftrightarrow{AB}$ and $\overleftrightarrow{AC}$ as shown.

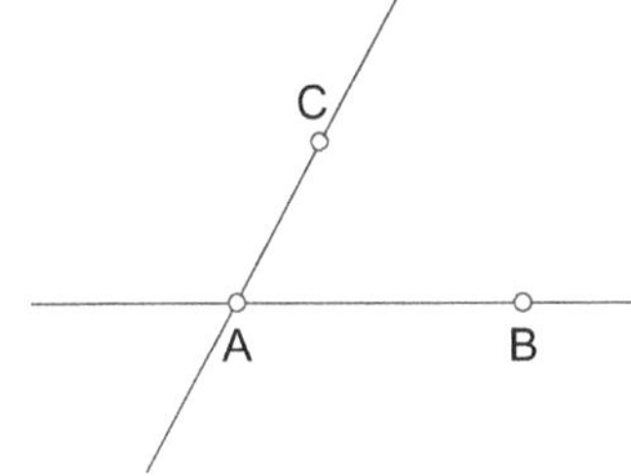

20. Label the points *A* through *C*.

21. Select, in order, points *C*, *A*, and *B* and choose **Transform | Mark Angle.**

22. Double-click point *C* to mark it as a center for rotation.

23. Select $\overleftrightarrow{AC}$. Then choose **Transform | Rotate** and click **Rotate.**

24. Drag the points of your sketch. Do the lines stay parallel? Explain why this method works.

Squaring the Sides: The Pythagorean Theorem

ACTIVITY NOTES

INTRODUCE

Project the sketch for viewing by the class. Expect to spend about 10 minutes.

1. Open Sketchpad and enlarge the document window so it fills most of the screen. If you are short on time, you can have students use **Squaring the Sides.gsp,** which already contains the **Square** custom tool that students create in worksheet steps 2–9.

2. Explain, ***Today you're going to use Sketchpad to explore the Pythagorean Theorem. You will create a custom tool that constructs squares and use it to construct squares on the sides of a right triangle. Then you'll measure lengths of the sides of the right triangle and the areas of the attached squares and make a conjecture about their relationship. Before you start, I'll model how to make the custom tool.*** As you demonstrate, make lines thick and labels large for visibility. Model how to create the **Square** custom tool. Here are some tips.

   - In worksheet step 1, model how to set the Sketchpad Preferences so that points are automatically labeled. ***For this sketch, we'll have Sketchpad automatically label the points as we construct them. It'll be easier to follow the construction.***

   - In worksheet step 3, explain to students that the marked point will flash briefly to indicate it has been marked. ***Point A will be used as a center for the rotation.*** Tell students that in the rotated image, the point corresponding to point B is B' and is read "B prime." ***What type of angle is $\angle B'AB$? How do you know?*** [It is a right angle because $\overline{AB}$ was rotated 90°.]

   - In worksheet step 4, model the rotation and again ask students to name the type of angle formed. ***What type of angle is $\angle AB'A'$?*** [Right angle]

   - In worksheet step 5, it may be easier to construct the segment by selecting point A' and point B and choosing **Construct | Segment.**

   - In worksheet step 6, remind students that squares are named by listing their vertices in consecutive order. ***What is this shape called?*** [Square $ABA'B'$ or square $BAB'A'$]

   - In worksheet step 7, model how to drag each vertex to test the construction. ***It is important not to skip this step. You need to test your construction to make sure it holds together.*** Model how to choose **Edit | Undo** if students make mistakes.

- In worksheet step 9, model how to press and hold the **Custom** tool icon to display the Custom Tools menu. When you are done naming the custom tool, press and hold the **Custom** tool icon to show students how the **Square** tool now appears in the Custom Tools menu. ***You can now use the Square custom tool to make other squares.*** Show students how to select and use the **Square** tool.
- You might also model how to add a new blank page to a sketch, as described in worksheet step 13.

3. If you want students to save their work, demonstrate choosing **File | Save As,** and let them know how to name and where to save their files.

DEVELOP

Expect students at computers to spend about 30 minutes.

4. Assign students to computers. Distribute the worksheet. Tell students to work through step 29 and do the Explore More if they have time. Encourage students to ask their neighbors for help if they are having difficulty with the construction.
5. Let pairs work at their own pace. As you circulate, here are some things to notice.
   - In worksheet step 10, listen to students as they discuss the properties of the square they used in the construction. Students should make the following points.

     When we rotated the segment by 90 degrees, we were making right angles.

     We were making an image of the side, so the sides are congruent.
   - In worksheet step 20, students may say that a right angle has a right, or 90°, angle. ***How do you know it has a 90-degree angle?*** [Perpendicular lines form 90° angles.]
   - In worksheet steps 21 and 22, check to be sure that students label their shapes as shown in the illustration. Explain that this will match the way the Pythagorean Theorem is usually stated and will make it easier for students to remember it. Students can use the **Text** tool to reposition labels, if needed.

- In worksheet step 23, some students will have difficulty attaching the squares to the vertices of the triangle. Suggest that students try attaching the right vertex first and then the left vertex.
- In worksheet steps 27–29, listen to students as they work on these steps. If students are having difficulty seeing any relationships, suggest that they use the **Arrow** tool to drag the measurements so that the side lengths are listed vertically in order from side length *a* to side length *c*, and the area of a square is next to the length of its attached side. You can also ask some pointed questions. ***Do you see a relationship between the two smaller square areas and the largest square area? How is a side length related to the area of its square?***
- If students have time for the Explore More, they will investigate whether the converse of the Pythagorean Theorem is true. Have students try finding more than one triangle where the sum of the two areas is equal to the third area. ***Can you make a good conjecture based on one triangle? Can you find another triangle where the sum of the two areas is equal to the third area? What type of triangle is this one?***

6. If students will save their work, remind them where to save it now.

SUMMARIZE

Project the sketch. Expect to spend about 5 minutes.

7. Gather the class. Students should have their worksheets with them. Open **Squaring the Sides Present.gsp** and go to page "Pythagorean Theorem."
8. Begin the discussion by reviewing the definition of a right triangle. ***What is a right triangle?*** Write "A right triangle is" on chart paper. Work with students to write a definition. Here is a sample definition: A right triangle is a triangle that has one right angle, or a 90° angle. ***How do you know the triangle you constructed was a right triangle?*** [It was constructed with perpendicular sides; perpendicular lines form right angles.]
9. ***What is the side opposite the right angle called?*** [Hypotenuse] ***What are the other sides called?*** [Legs] ***In the sketch, which side is the hypotenuse and which sides are the legs?*** [The hypotenuse is side *c*; the legs are sides *a* and *b*.] Drag the right triangle around so students can see the right triangle in many positions and in many sizes. Point out that

the longest side is always the hypotenuse and the right angle is always opposite it.

10. Review how to find the area of a square. ***What is the formula for area of a square?*** [$A = s \times s$, or s^2] ***How would you find the area of any size red square in the sketch?*** [$A = b \times b$, or b^2] Have students state how they would find the areas of the other two squares as well.

11. Drag the vertices of the triangle. ***Did you find a relationship among the three areas? Explain.*** Students may make the following statements.

    Yes, we noticed that the sum of the areas of the two smaller squares equals the area of the largest square.

    We found the same thing, but we said it differently. The area of the largest square minus the area of the smallest square equals the area of the mid-sized square.

    Let's check your reasoning. Press *Show Calculation.* Drag the triangle vertices again to show that no matter the size of the right triangle, the sum of the areas of the two smaller squares equals the area of the largest square.

12. ***Based on your observations, what equation did you write that relates side lengths a, b, and c in any right triangle? Explain your thinking.*** Students should reason that the area of the square with side length a is a^2; the area of the square with side length b is b^2; and the area of the square with side length c is c^2. The sum of the areas of the two smaller squares equals the area of the largest square, or $a^2 + b^2 = c^2$. Tell students this is known as the Pythagorean Theorem and it holds true for any right triangle with leg lengths a and b and hypotenuse length c.

13. If time permits, discuss the Explore More. ***What did you discover?*** Students should find that the converse of the Pythagorean Theorem holds true: If the area of the largest square is equal to the sum of the areas of the smaller squares, then the triangle is a right triangle.

14. ***A right triangle has legs with lengths of 6 cm and 8 cm. How could you find the length of the hypotenuse?*** [Take the square root of the sum of the squares of the sides: $6^2 + 8^2 = c^2$; $\sqrt{6^2 + 8^2} = \sqrt{c^2}$; $\sqrt{36 + 64} = c$; $\sqrt{100} = c$; $10 = c$]

EXTEND

You might extend this activity by asking students whether they think the relationship among the areas holds for other similar shapes built along the sides of the right triangle. ***If you constructed equilateral triangles along the sides of the right triangle, do you think the area of the largest triangle would equal the sum of the areas of the smaller triangles?*** [The areas of the regions built on the legs of a right triangle will always total the area of the region built on the hypotenuse as long as all three regions are similar.] If you have time, you can model this fact.

ANSWERS

10. This construction uses the following property: A square has right angles and congruent sides.

20. This construction uses the following property: A right triangle has one side perpendicular to another side.

28. The sum of the areas of the two smaller squares equals the area of the largest square.

29. $a^2 + b^2 = c^2$

    Students already familiar with the Pythagorean Theorem are likely to write this when they make a conjecture. You might also have them express the theorem in words.

30. If the sum of the areas of squares on two sides of a triangle equals the area of the square on the third side, the triangle must be a right triangle.

Squaring the Sides

Name:

In this activity you'll construct squares on the sides of a right triangle using a custom tool. Then you'll use the areas of these squares to explore perhaps the most famous relationship in mathematics—the Pythagorean Theorem.

CONSTRUCT

1. In a new sketch, choose **Edit | Preferences.** On the Text panel, under **Show labels automatically**, check **For all new points.** Click **OK.**

In steps 2–9, you will create a custom tool that constructs a square.

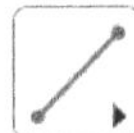

2. Construct $\overline{AB}$.

3. You'll mark point *A* as a center and rotate point *B* and $\overline{AB}$ by 90°.
   Double-click point *A* to mark it as a center.
   Select point *B* and $\overline{AB}$. Then choose **Transform | Rotate.**
   In the Rotate dialog box, select **Fixed Angle** and enter 90 for degrees.
   Click **Rotate.**

4. Mark point *B'* as a center and rotate point *A* and $\overline{B'A}$ by 90°.

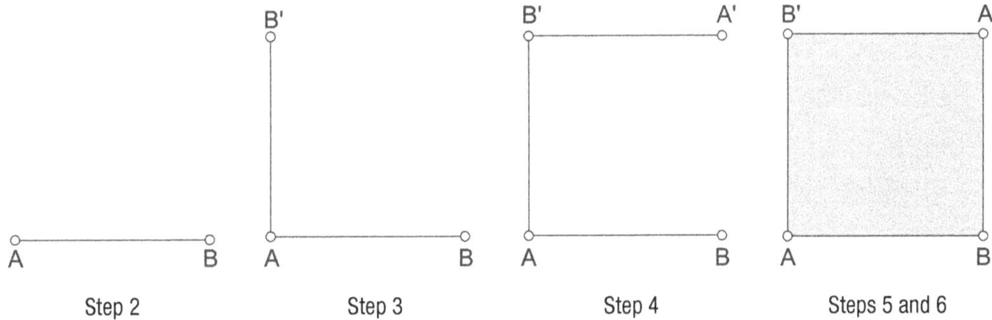

5. Construct $\overline{A'B}$ to finish the square.

6. Select the vertices in order and choose **Construct | Quadrilateral Interior.**

7. Drag each vertex of the square to make sure it holds together.

8. Hide the labels by clicking on each point.

9. Now you'll make a custom tool of this construction that you can use to make other squares.

Select the entire figure. One way is to choose **Edit | Select All.**

Choose **Create New Tool** from the Custom Tools menu.

In the Tool Box dialog box, enter Square for the Tool Name.

EXPLORE

10. What properties of a square did you use in this construction?

11. Choose **Edit | Preferences.** On the Text panel, under **Show labels automatically,** uncheck **For all new points.** Click **OK.**

12. Use the custom tool to get a feel for the way it works. Note that the direction in which the square is constructed depends on how you use the tool.

CONSTRUCT

13. Now you'll add a new page to your sketch.

Choose **File | Document Options.**

In the Document Options dialog box, select Add Page and choose **Blank Page.** Click **OK.**

In steps 14–18, you'll construct a right triangle. Then in step 23 you'll use your **Square** tool to attach squares onto each side of the right triangle.

14. Construct $\overline{AB}$. Label the points if you find it helpful, but you'll be changing the labels in step 21.

15. Select point A and $\overline{AB}$ and choose **Construct | Perpendicular Line.**

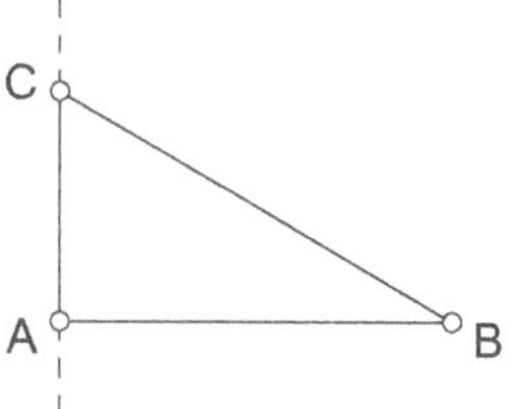

16. Construct point C on the perpendicular line.

17. Select $\overleftrightarrow{AC}$ and choose **Display | Hide Perpendicular Line.**

18. Construct $\overline{AC}$ and $\overline{BC}$.

Squaring the Sides

continued

19. Drag each vertex to confirm that your triangle stays a right triangle.

20. What property of a right triangle did you use in your construction?

21. Click on the vertices in order from *A* through *C* as shown at right, so that *C* is the right-angle vertex.

    To change a label, double-click it, enter the new label, and click **OK.**

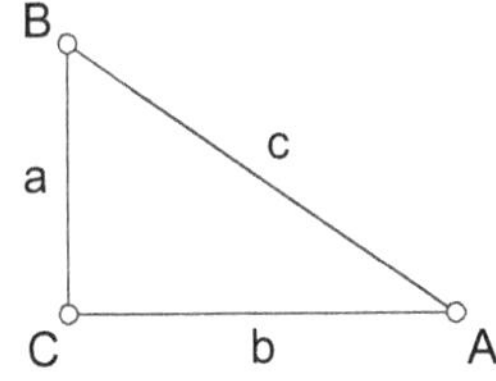

22. Show the labels of the sides by clicking on each segment. Change the labels to *a, b,* and *c* so that side *a* is opposite ∠*A,* side *b* is opposite ∠*B,* and side *c* is opposite ∠*C.*

23. Use your new **Square** custom tool to construct squares on the sides of your triangle. Be sure to attach each square to a pair of the triangle's vertices. If your square goes the wrong way (overlaps the interior of your triangle) or is not attached properly, choose **Edit | Undo** and try attaching the square to the triangle's vertices in the opposite order.

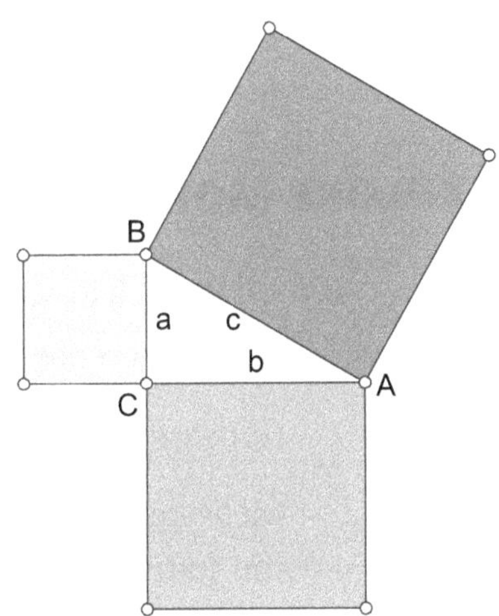

24. Drag the vertices of the triangle to make sure the squares are properly attached.

EXPLORE

25. Measure the areas of the three squares by selecting each interior and choosing **Measure | Area.**

26. Measure the lengths of sides *a, b,* and *c* by selecting each side and choosing **Measure | Length.**

27. Drag each vertex of the triangle and observe the measures.

28. Now you'll use the Calculator to write an expression based on your observations.

    Choose **Number | Calculate** to show the Sketchpad Calculator.

    Click once on a measurement to enter it into a calculation.

    Describe any relationship you see among the three areas.

29. Based on your observations about the areas of the squares, write an equation that relates *a*, *b*, and *c* in any right triangle. (*Hint:* What's the area of the square with side length *c*? What are the areas of the squares with side lengths *a* and *b*? How are these areas related?)

EXPLORE MORE

30. Now you'll investigate the converse of the Pythagorean Theorem.

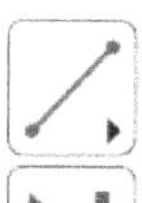

    Construct a generic triangle (not a right triangle).

    Construct a square on each side.

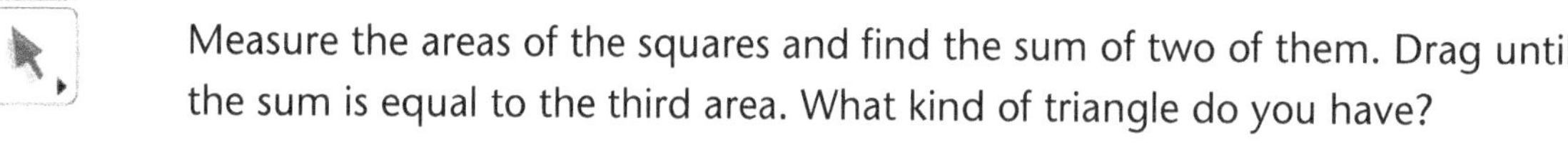

    Measure the areas of the squares and find the sum of two of them. Drag until the sum is equal to the third area. What kind of triangle do you have?

Apartment Floor Plan: Scale Drawing

INTRODUCE

Be ready to project the sketch for viewing by the class in step 2. Expect to spend about 45 minutes, the first class period.

1. ***Imagine there is an old warehouse in your neighborhood that is being made into apartments. You've been hired to design the apartments. Like real architects, first you're going to make a rough*** **floor plan** ***using paper and pencil. Can anyone explain what a floor plan is?*** Bring out the idea that a floor plan is a view of a building's interior that shows everything from above, as though the ceiling were gone and you were looking straight down on the rooms. Students may also share ideas about scale.

2. Open **Apartment Floor Plan Present.gsp** and go to page "Example." ***You'll use your rough floor plan to make an accurate floor plan like this one, called a*** **scale drawing,** ***using a grid in Sketchpad. Because architects' floor plans can't be the actual size of buildings, architects*** **scale down** ***the measurements (or*** **dimensions*****) when they make scale drawings.*** Read the scale on the drawing: 1 centimeter represents 1 meter. Make sure students recognize the abbreviations for centimeter and meter.

   Make groups of three to four students and distribute at least one meterstick to each group. Have the groups locate a distance of 1 centimeter on the meterstick and compare it to 1 meter. Elicit the observations that a meter is 100 times as large as a centimeter: 100 centimeters equal 1 meter.

When writing a scale, the drawing's dimension is written first, followed by the real-life dimension.

3. ***In a scale drawing you show the floor area. How do we measure area?*** [In square units] If your classroom floor is covered with square tiles, point them out.

   Go to page "Area." ***Using the scale for the apartment floor plan—1 centimeter represents 1 meter—what will 1 square centimeter on your floor plan represent?*** [1 square meter in the actual apartment]

   Have the groups construct the outline of a square meter on the floor, using one or more metersticks. If groups have fewer than four metersticks, have students use tape to mark the outline or the corners of the square meter.

4. Give each group a pre-cut square centimeter to place inside their square meter. ***How many square centimeters do you think it would take to fill up the inside of this square meter?***

Sample Class Discussion

Jason: *I think it would take 100 because there are 100 centimeters in a meter.*

Margo: *There have to be more. One hundred of those little squares would fill up just one row all the way across the square meter!*

Susan: *There would have to be 100 rows of 100.*

Sonny: *What's 100 times 100? Four zeros after a 1—that's 10,000. It would take 10,000 square centimeters to cover the square meter.*

Teacher: ***Every square meter of the apartment you design will be shown as a square centimeter in your scale drawing.*** (Teacher provides time for reflection on this fact to help students understand that the scale drawing will be much smaller than the real apartment.)

Consider partnering students who have strong computation skills with students who are not as strong.

5. Partner students and have them sit together. Go back to page "Example." ***You and your partner will develop a design together. Your apartment design may be quite different from this one, but it has to follow these design rules set down by the building owner.***

- The apartment must be rectangular and have a total area of 100 square meters.
- All rooms and other spaces must have square corners.
- One side of the apartment must have windows. The other three sides will not have windows.
- There must be two bedrooms.
- All dimensions must be in whole meters.

Talk with your partner about the apartment shown in this plan. Exchange ideas about what you like and what you might change. Explain these features of the plan: The label C indicates a closet; the label W indicates a window; dashed segments at closets represent sliding doors; and doors that swing are shown with the symbol that includes the green arc.

Give students a few minutes and then facilitate a brief class discussion about the features of the apartment. Call attention to the traffic paths, something students may not have considered.

Draw attention to the grid in order to discuss how students will calculate the area of rooms and other spaces. ***How big is the kitchen? How did***

you decide that? How big is the living room/dining area? How did you find that? Some students are likely to have found the areas by multiplying the dimensions; others may have counted squares. Offer multiplying as a faster method, but accept the counting strategy as a valid alternative.

How big is bedroom 1, without the closet? How did you find that? Because this is not a rectangular space, students' strategies may vary. Elicit a variety of methods. Some students may have partitioned the space into a rectangle with "extra" squares. Others may have counted squares. And others may have imagined a 3-by-5 rectangle, multiplied, and then subtracted the one square that is part of the bathroom.

6. ***Let's create another apartment design.*** Go to page "Do Your Own." ***I'm going to model what you will do when you make your scale drawing.***

Some students may question whether a square is a rectangle. Don't let this sidetrack the activity. Offer a definition and move on.

We have to follow the design requirements for this apartment. It has to be exactly 100 square meters. And it has to be rectangular. One way to get 100 square meters is to make the dimensions of the apartment 10 meters by 10 meters.

What other dimensions would give us an area of 100 square meters? What are some other factors of 100? As students make suggestions, develop the idea that some factor pairs would generate dimensions that result in better apartment shapes than others. Introduce the important idea that the practical choices available to students as they are designing will be fewer than the mathematically possible choices. Work with different dimensions to encourage students to think about the options. One reasonable choice is 5 m by 20 m.

7. As you draw the layout and traffic paths of the apartment, think aloud and invite student comments and questions. Follow these steps.

   - Construct the perimeter of the apartment, using the **Segment** tool and choosing **Display | Line Style | Thick.** Explain that **Graph | Snap Points** is chosen so that segments align with the grid and walls will measure in whole centimeters.

   - Place one or more windows along one side of the apartment. Explain that for the sake of simplicity, students will show windows as segments in a color different from the color used to show walls. (Architects use another symbol.)

- Locate the entrance and demonstrate how to draw doors. Because it takes a few steps to draw a door, consider making a large door to one side of the floor plan as an example.

  Start with two points (*A* and *B* in the illustration here) that mark the location and width of the door. Draw a segment (*AC* here) from one of the points to indicate the door and the direction it opens. (Point labels here are for descriptive purposes only; students won't be labeling points.)

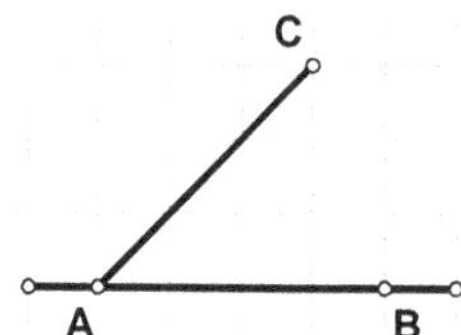

Note that you can draw a point that does *not* snap to the grid by placing the point away from the grid intersections.

Draw a point approximately midway along the door opening. Select sequentially the three points indicating the door opening and choose **Construct | Arc through 3 Points.**

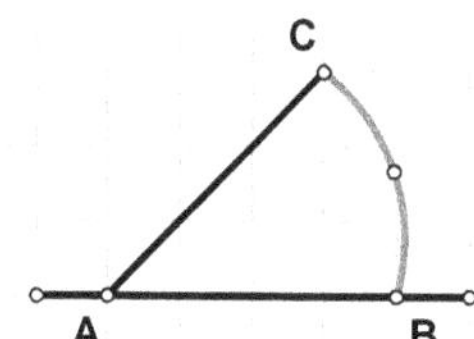

Hide the middle point of the arc by selecting it and choosing **Display | Hide Point.** Hide the original wall segment, draw new segments on either side of the door, and use **Display | Line Style | Dashed** for a segment you draw across the doorway itself.

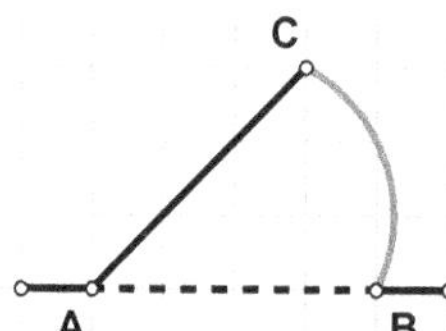

Remind students that all dimensions and area measurements must be whole unit amounts.

- Partition the interior into rooms, drawing walls in short segments that can be easily changed by dragging, rather than in long segments that span several rooms.
- Have the class mentally compute the area of each room, as you create it, and give an equation for finding the area.
- Using the **Text** tool, create captions to label the rooms.

- Demonstrate choosing **File | Save As,** and let students know how to name and where to save their files.

Emphasize that the apartment students design need not match their own home.

8. ***Before you design an apartment, you will measure spaces at home to gather information that may guide you in making decisions. For example, what dimensions of bedrooms allow the furniture to fit well and people to move easily around the room? In which direction do doors open? How much light do large windows let in? Small windows?*** Distribute the first page of the worksheet and the metric measuring tapes or 1-meter paper strips students will take home to use.

   Have two volunteers measure a wall of the classroom using a metric measuring tape or meterstick. ***About how many meters long is this wall of our classroom?*** Point out that construction in students' homes may be based on feet, not meters, so students may need to think in approximate terms as they look at the size of spaces in their own homes.

   Distribute measuring tapes for use at home.

DEVELOP

Expect students to spend about 30 minutes on the rough floor plan and 30 minutes at computers creating the scale drawing.

9. On a following day, after students have collected measurements at home, seat students with their partners. Students should have the first page of the worksheet with the measurements they recorded at home. Distribute blank paper and the remaining pages of the worksheet. Read the *Rough Floor Plan* section of the worksheet (step 2) with the class. Tell students to aim to complete their floor plan in 30 minutes.

   As you observe partners working, check whether they are incorporating all the requirements. ***Have you included all the information you will need for your scale drawing? When you have a design, try to "walk" through the apartment. If you want to make changes, do.***

10. When students are ready to move to computers, tell them where to locate **Apartment Floor Plan.gsp** and let them know they will need their worksheets. ***Do steps 3 through 8. If you have time, do the Explore More.*** Encourage students to ask a neighbor for help if they have questions about using Sketchpad.

11. Let pairs work at their own pace. As you circulate, listen to students' conversations and observe their designs in order to learn about their thinking. Note whether students are showing their area calculations as

multiplication equations. Accept addition equations only from those students for whom that is cognitively appropriate.

Be sensitive to differences in students' real-world housing as you remind them that this is their chance to design a space that will not have problems that many homes have.

12. Let students know when about 10 minutes are left so they can plan to bring their work to an end. Students who have time can do the Explore More. Remind students how to name and where to save their files.

13. If a flash drive is available, collect student sketches to display on the shared computer during the Summarize discussion. Alternatively, plan time to print out the sketches for display. Some floor plans may require printing on more than one page in order for the drawing to remain to scale. Alternatively, students could choose **File | Print Preview**; in the dialog box that appears, they would set the image to fit on one page before clicking **Print.** Explain to students that the unit distances, however, will be less than a centimeter.

SUMMARIZE

Project students' sketches. Expect to spend 15 minutes. Consider printing and displaying any floor plans that you do not have time to discuss.

14. Gather the class and facilitate discussion of students' experiences and thinking. Here are questions the discussion may address.
    - How did students decide on the size of the doorways?
    - Did students think about furniture in the rooms and where it might go?
    - How did students' measurements at home help them make decisions about the apartment?
    - Did using the metersticks to see actual distances cause students to change or confirm dimensions used in their scale drawing? How?
    - Is the scale the class used a good scale? Why or why not?
    - What would be a good scale for working with customary U.S. measures—yards, feet, and inches?

15. ***What other kinds of things have you seen that are scale drawings or scale models?*** Expect students to suggest such things as model cars and trains, dollhouses, and maps. Offer your own ideas only if students have difficulty thinking of things. The goal is to prompt students to realize that scale drawings and models are all around them.

16. If time permits and students had time to do the Explore More activity, ask volunteers to share their results.

EXTEND

Have students create a scale drawing that shows an object at a size that is larger than life. Gather the class. Facilitate discussion of times when a scale drawing that is *larger* than the actual thing is useful.

With the class, read the *Larger Than Life* Steps 10–14 on the worksheet. Project **Apartment Floor Plan.gsp.** Go to page "Pencil." Point out these features of the sketch: the axes, but not the grid, are visible; **Graph | Snap Points** is not chosen (demonstrate using the **Point** tool); and units are set to tenths of a centimeter (demonstrate by drawing a segment and choosing **Measure | Length**).

Students should again work in pairs. As you observe, ask them to compare their scale drawings to the actual pencil to verify that the drawings look to be in proportion. ***How many times the actual size of the pencil is your drawing?*** [Two times]

ANSWERS

1. Measurements will vary.

8. Measurements will vary. The total of all areas should be 100 square meters. Here is a sample, using the dimensions of the example floor plan.

Room	Equation	Area (m²)
Bedroom 1	$3 \times 4 + 1 \times 2 = 14$	14
Bedroom 2	$3 \times 6 = 18$	18
Bathroom	$2 \times 3 + 2 = 8$	8
Kitchen	$3 \times 4 = 12$	12
Closets (all)	$1 \times 2 + 2 \times 1 = 4$	4
Entry and Hall	$2 \times 4 + 2 \times 3 + 2 = 16$	16
Dining and Living Room Areas	$4 \times 7 = 28$	28
TOTAL		100

Apartment Floor Plan

Name:

EXPLORE

1. At home, take measurements using a meterstick or tape measure. Round the measurements to the nearest meter. Record in the table.

Home Measurements

Room or Space	Length (m)	Width (m)	Area (m^2)
Door 1			
Door 2			
Hall			

ROUGH FLOOR PLAN

2. On blank paper, draw a rough floor plan with your partner. You may use metersticks to decide the size a space should be. Think about the measurements you made at home too.

   Remember, the owner of the building has these requirements.

   - The apartment must be rectangular.
   - All rooms and other spaces must have square corners.
   - One side of the apartment should have windows. The other three sides will not have windows.
   - There must be two bedrooms.
   - The total area of the apartment must be exactly 100 square meters.
   - All dimensions must be whole meters.

SCALE DRAWING

Now you will use the information in your rough floor plan to make a scale drawing of your apartment design.

3. Open **Apartment Floor Plan.gsp.** Go to page "Scale Drawing." Enlarge the document window so it fills most of the screen.

4. At the top of the page, type a caption with your names. Below, type a caption with the scale (1 cm represents 1 m).

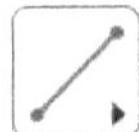

5. Start to draw the perimeter of the apartment. When you have drawn the first segment, while it is selected, choose **Display | Line Style | Thick.**

   Show windows, using segments in a different color. Select one or more segments, choose **Display | Color,** and pick a color.

6. Draw the rooms and other spaces, such as closets and hallways. *Tip:* Do not draw a wall longer than a side of one room.

   To include doors, follow these steps.

   - Start with two points that mark the doorway. Use the **Point** tool if you need to make the points.

- Draw a segment showing the direction the door opens.
- Add a point, select three points, and choose **Construct | Arc through 3 Points.**

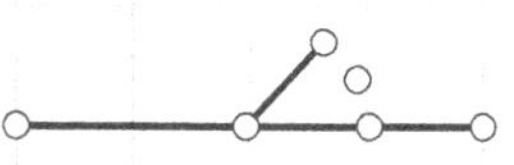

- Select the middle of the three points and choose **Hide | Point.**
- Make a segment across the doorway and choose **Display | Line Style | Dashed.** (If needed, select a wall segment and choose **Hide | Segment**; then draw wall segments on each side of the door.)

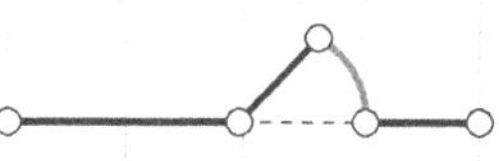

7. Label the rooms and other spaces with their names.

8. Use mental math to find the area of each room and other space. Write your calculations as equations in the table.

Area of Rooms

Room	Equation	Area (m^2)
Bedroom 1		
Bedroom 2		
Bathroom		

EXPLORE MORE

9. Draw a stick figure next to your floor plan. The stick figure should represent a basketball player who is two meters tall.

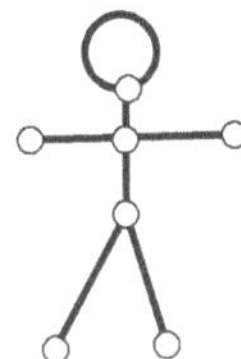

LARGER THAN LIFE

10. Now use scale drawing to make a big drawing of a small thing. Find a sharpened pencil that is between 8 cm and 15 cm long.

11. Go to page "Pencil." At the top, type your names and the scale 2 cm represents 1 cm. This means that every centimeter of pencil length will be shown as 2 cm in your sketch.

12. Measure all parts of the pencil carefully. Round measurements to the nearest tenth of a centimeter (nearest millimeter). Record in the table. Then record the measurements you will use for the scale drawing (2 times the actual measurements).

Pencil Measurements

Part of the Pencil (see the picture above)	Actual Measurement (rounded to nearest mm)	Scale Drawing Measurement
Entire pencil length		
Width of pencil		
Eraser (1)		
Metal band (2)		
Main body of pencil (3)		
Sharpened wood area (4)		
Tip (5)		

13. Make your scale drawing. Use the distances on the axes to help you draw segments that are the correct length. To check a segment's length, select the segment and choose **Measure | Length.** Change the segment, if needed.

14. Now you will color your drawing. In a clockwise or counterclockwise direction, select, in order, the points that surround a region. Choose **Construct | [Polygon] Interior.** With the interior selected, choose **Display | Color** and pick a color.

INTRODUCE

Project the sketch for viewing by the class. Expect to spend about 30 minutes.

1. Present the following situation. ***A class is planning to make popcorn balls for sports day at their school. The class that made the popcorn balls last year says that one cup of raw (uncooked) popcorn will make four popcorn balls.***

Use a Data Table

2. ***The students in the club started this table. Let's complete it.*** Show the recording of this partially filled table, which you have prepared on chart paper or a transparency. Have students explain their reasoning as the class completes the table.

Raw Popcorn (in cups)	Popcorn Balls
1	4
2	
4	
8	

3. ***What patterns do you see in the table?*** The columns in the data table show that the number of popcorn balls doubles as the number of cups of raw popcorn doubles. The rows in the data table show that the number of popcorn balls is always four times the number of cups of raw popcorn.

Choose a Type of Graph

4. Tell students that some of the club members wonder whether making a graph might help them answer questions that will come up. For example, they know that last year 102 popcorn balls were sold. How many cups of raw popcorn, they wonder, would be needed to make that many popcorn balls? ***What would be the best kind of graph for the club to use?*** The discussion should bring out the ideas that a point graph is often used when both variables are numbers and when you want to know about values between the data points.

5. Open **Point Graphs.gsp** and go to page "Popcorn." Explain that the class will learn to use Sketchpad to create a point graph.

If the class can justify the use of the scales on a different page of the sketch, use that page rather than page "Popcorn."

Label the Axes

6. Discuss which variable should be graphed on each axis. We'll assume here that the class graphs the values from the first column of the data table on the horizontal axis, and the second column on the vertical axis. Refrain from labeling the axes until step 8 below.

7. Discuss the scales of the axes. Students should note that the horizontal and vertical axes are scaled differently. ***Will the numbers on the axes work well for what we want to graph?*** Go to several other pages and discuss the numbers on the axes. ***Would these be more appropriate?*** Elicit students' thinking about the range of the data, the appropriateness of the intervals shown on the axes, and room for *extrapolation* (if students are familiar with making predictions for values that go beyond the existing data).

   Students may suggest that they will be able to scroll to see more of the graph if necessary. Confirm that when they have created their point graphs and a line through the data, they will be able to scroll in their sketches. Note that you can adjust the scale of an axis by dragging the tick labels. This is another way to see more of the graph on screen.

It is not possible to rotate text 90° to label the *y*-axis.

8. When the class has decided on the scales to use, use the **Text** tool to label the axes by creating a caption with the name of the variable for each axis. (You may want to add a letter to stand for the variable.) Ask the class to suggest a title for the graph. Create a caption containing the title near the top of the screen.

Plot Points

9. Model graphing the data in the data table. Choose **Graph | Plot Points.** In the Plot Points dialog box that appears, enter the coordinates for the first point you want to plot, (1, 4), and click **Plot.** Repeat to plot the remaining data in the table. Discuss including (0, 0) as a data point and have students observe that there is already a point at the origin of the Sketchpad grid. When all points are plotted, click **Done.**

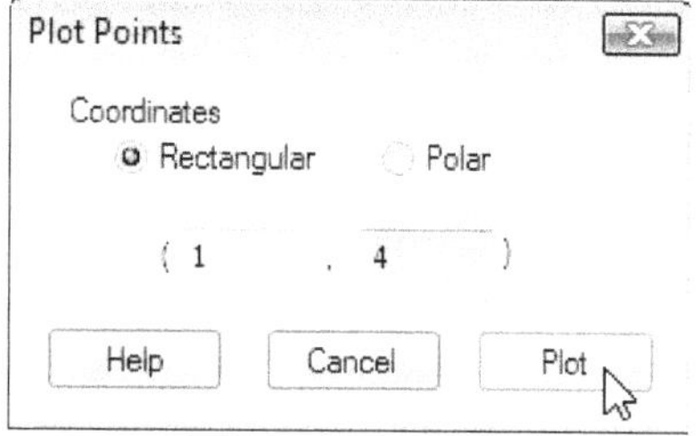

10. Using the **Arrow** tool, select all of the data points. Choose **Display | Color** and choose a color other than red. (This will help students distinguish the data points from other points in the graph.) Make sure students can see the points in the new color.

Look for Patterns

11. ***What do you notice about these points?*** Students should observe that each point is "over and up" the same amount from the previous point, that the points go up and to the right, and that the points lie along a line.

    Let's fit a line to the points. Using the **Line** tool, click in empty space on the grid to locate a first point through which the line passes. Click again in empty space to locate a second point through which the line travels.

    Invite a volunteer to adjust the line by dragging the red points. Let students confirm that the line goes through all the data points; the points lie exactly on a straight line. Point out that the point at (0, 0) also lies on the line. ***Does that make sense? Why or why not?*** A sample student explanation is this: *If no popcorn is used, then no popcorn balls are made.*

Predict

12. Pose this question: ***One of the club members said, "I want to know how much popcorn to use to make 10 popcorn balls."*** Have students suggest how to use the graph to answer the question.

13. Demonstrate how to interpolate using the dashed red lines at the left and bottom edges of the screen.

    - Drag the horizontal red line up until you reach 10 on the vertical axis (number of popcorn balls). Observe that the red line extends through the line through the data points.

    - Drag the vertical red line to the right until it meets the line at the point where the red horizontal line also meets it. Observe that the vertical red line extends through the horizontal axis (number of cups of popcorn).

    - With the class, estimate the value of the location where the red line meets the horizontal axis. [The location is exactly 2 1/2.]

Go back to the data table and ask students whether using 2 1/2 cups of popcorn for 10 popcorn balls makes sense. Here are samples of student reasoning.

It makes sense because 2 1/2 is halfway between 2 and 3, and 10 is halfway between 8 and 12.

From 8 popcorn balls, it takes a cup to make 4 more, so it only takes half a cup to make 2 more.

14. ***Can we figure out how many cups of popcorn would be needed to make 20 popcorn balls?*** Have students talk in pairs or small groups, and then as a class. Discussion should bring out that because 20 balls is twice 10 balls, twice 2 1/2 cups, or 5 cups, would be needed.

    Let's see whether we get the same answer using the graph. How can we use the red lines to check? Invite a volunteer to help you at the computer. The lines may be dragged in either order. For the sake of example, let's assume here that the vertical red line is dragged to 5 on the horizontal axis (number of cups of popcorn); now the horizontal red line can be dragged up to meet the vertical red line at the line through the data points. ***Where does the horizontal line cross the vertical axis (number of popcorn balls)?*** [At 20]

15. Invite the class to propose questions that can be answered using extrapolation. Choose a question and invite a volunteer to the computer to drag the red lines in order to use the graph to answer the question. The class should suggest scrolling to the right and/or up, if needed.

16. Model using **Edit | Save As** to rename the file and save it. This is a convenient stopping place if you want to do this activity over two days.

DEVELOP

Continue to project the sketch. Expect to spend about 20 minutes.

17. Go to page "Cookie Sales." ***This graph shows the money a class made when they sold cookies over eight days for a fundraiser.*** Ask students to describe the graph. ***What do you notice about these data points?*** Students should observe that the points are "around a line" and go in an uphill direction.

Create a Line that Approximates the Data

18. Construct a line in the grid as you did in the previous graph. Model dragging the two points on the line to move the line.

19. Have a volunteer come to the computer to position the line so that it appears to fit the data points as closely as possible. If students have other ideas about where the line should be positioned, invite them to the computer to position the line.

Predict

20. Have the class pose questions that can be solved by extrapolating. Invite volunteers to drag the red lines to make predictions in response to several of the class's questions.

SUMMARIZE

Project the sketch. Expect to spend about 10 minutes.

21. Distribute the worksheet so students can preview it for use at another time. Tell them that they will have a copy of the worksheet when they use Sketchpad in a graphing activity in the future. Give students time to read through the steps while today's demonstration is fresh in their minds.

22. Facilitate an exchange of ideas about the use of technology to create and analyze data. If students have used other software for data investigations, include discussion of those technologies as well.

EXTEND

1. Open **Point Graphs Present.gsp.** Go to page "Extend" to explore another way to interpolate and extrapolate. Drag point *A* and notice that its coordinates update. Use the displayed coordinates to find data points between and beyond the initial data points. You will want to discuss rounding with students when they see that Sketchpad, because of the way it's built, will sometimes display the same coordinates for two points that are near each other.

If students will use this method in another graph, give these directions.

- Using the **Point** tool, construct a point *A* on the line you have drawn through the data points (or, as appropriate, the line that fits the data points as closely as possible).
- With the point selected, choose **Measure | Coordinates.**
- Drag the point along the line. Use the displayed coordinates to find data points between and beyond the initial data points.

2. When using Sketchpad to create point graphs for data investigations, here are some additional options available to students.

   - If students will plot data points with values other than whole numbers, have them choose **Edit | Preferences** and, in the Units panel, set the precision for **Others** to **tenths** or **hundredths,** as appropriate.
   - Have students scale the axes themselves by dragging the unit point or tick labels on each axis.
   - Have students graph additional data sets on a graph following the directions in worksheet step 9.
   - When graphs are complete, have students prepare them for display. If students will print their graphs, make sure they choose **File | Print Preview** and, in the dialog box that appears, set the image to fit on one page and then click **Print.**

Point Graphs

Name:

Use Sketchpad to make a point graph.

LABEL THE AXES

1. You should have a table of data. Decide which variable goes on each axis of the graph you will create.

2. Open **Point Graphs.gsp.**
   Look through the pages to see the ways the axes are numbered. Choose a page for graphing your data.

3. Create captions to label each axis and title your graph.

PLOT POINTS

4. Now you will plot your data.
   Choose **Graph| Plot Points.**
   In the dialog box that appears, enter the coordinates for the first point you want to plot, and click **Plot.**
   Repeat to plot the rest of the data.
   When all points are plotted, click **Done.**

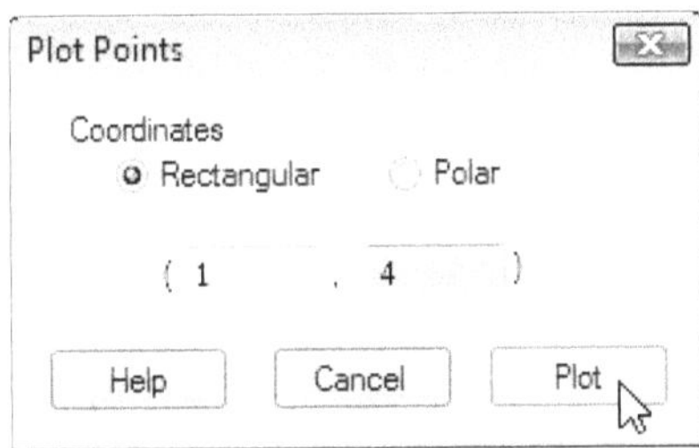

LOOK FOR PATTERNS

5. Do the data points lie on a line? If they do, draw a line going through all the points.
   Does the point at (0, 0) lie on the line? Decide whether it makes sense to include that point as a data point.

6. If the data points do not lie on a line, do they lie close to a line? If so, draw a line and then, using the **Arrow** tool, drag the points on the line so that the line fits the data points as closely as possible.

PREDICT

7. Use the dashed red lines to find the locations of data points other than those you plotted.

8. Scroll up and/or to the right if you need to see more of the graph.

OTHER DATA SETS

9. Do you have another set of data to plot on the same graph? If so, follow steps 3–6.

   For each set of data you plot, use a new color for the data points and for the graphed line or line of best fit.

   Create a caption to label each set of data.

Dartboards: Geometric Probability

INTRODUCE

Project the sketch on a large-screen display. Expect to spend about 10 minutes.

1. Explain, ***Today we're going to throw darts—imaginary darts—in Sketchpad.*** Open the document **Dartboards.gsp** and go to page "First Throws." Press *Throw Dart Slowly* several times to demonstrate how the dart hits the rectangle randomly.
2. ***Let's make a target within the rectangle.*** Use the **Segment** tool to draw a polygon inside the rectangle. Show how to construct the polygon's interior and color it, say, red.
3. ***Now let's throw the dart. Count how many times it hits the target.*** Make a class decision before you begin about whether darts landing on an edge of a target will count as hitting the target. Have a volunteer press *Throw Dart Slowly* 20 times. Resist student pleas to throw the dart fast, saying that they can throw the dart at whatever speed they wish when working on their own. (This will make it hard to keep an accurate count of hits.)
4. If you want students to save their work, demonstrate choosing **File | Save As,** and let them know how to name and where to save their files.

DEVELOP

Expect students at computers to spend about 25 minutes.

5. Distribute the worksheet. Assign students to computers and tell them where to locate **Dartboards.gsp.** Tell students to work through step 16 of the worksheet and do the Explore More if they have time. Encourage students to ask a neighbor for any help needed using Sketchpad.
6. Let students work at their own pace. As you circulate, here are some things to notice.
   - In worksheet step 3, some students may make their targets bigger than the board to guarantee that all darts will hit.
   - Students might have trouble with worksheet steps 10–14. If you see many students become frustrated you might want to bring the class together and model this process. Make sure you have practiced this process beforehand.
   - Although most students will see a connection between the relative areas and the portion of throws that hit the target, they may not be able to articulate the relationship very clearly yet.

7. If students will save their work, remind them where to save it now.

SUMMARIZE

Expect to spend about 10 minutes.

8. Gather the class and ask selected students to share their ideas. ***Are the areas related to the number of hits?*** Try to elicit a clear statement of the relationship. This statement might refer to ratios, fractions, or percentages. Here are some possibilities.

   The percentage of hits is the same as the percentage of target area to total area of rectangle.

   The ratio of the smaller area to the larger one is the same as the ratio of hits to throws.

   The target's area is a certain fraction of the rectangle's area; that fraction is the same as the fraction of throws that will hit the target over the total throws, or $\frac{\text{target area}}{\text{total area}} \approx \frac{\text{hits}}{\text{throws}}$.

9. Review or introduce the term *probability* as a synonym for this ratio. ***Another word for the ratio of hits to throws is probability. If*** $\frac{1}{3}$ ***of the throws are hits, we say that the probability is*** $\frac{1}{3}$ ***that a throw will be a hit.***

10. Discuss the two kinds of probability. The ratio of areas is the *theoretical probability* that a throw will be a hit. The ratio of actual results is the *experimental probability.* ***How do these two kinds of probability relate? How are they different?*** Bring out the idea that the experimental probability gets closer to the theoretical probability as the number of throws increases. Students might notice that in just a few throws, they sometimes hit the theoretical ratio. For large numbers, the experimental ratio is likely to be closer to the theoretical, but it is *less* likely to be *exactly* equal to the experimental ratio.

11. ***What other questions can you ask about the dart throwing?*** Encourage all student inquiry. You might list these questions of mathematical interest.

    What if the target consists of more than one shape?

    Why is ratio of areas about the same as the ratio of hits?

Why aren't those two ratios always exactly the same?

Why does experimental probability get closer to the theoretical probability with more and more throws?

How do you calculate probabilities for throws of a coin? For a pair of dice?

How do you calculate other probabilities, such as the chance of rain?

ANSWERS

3.–6. Answers will vary.

7. The ratio of the target area to the rectangle area.

8. Answers will vary.

9. Answers will vary, but should be about 10 times; you might find the mean for the class—total hits divided by total throws.

Dartboards

Name:

You'll see what you can learn about shapes from throwing Sketchpad darts at them.

EXPLORE

1. Open **Dartboards.gsp** and go to page "First Throws." Use segments to create a target polygon inside the rectangle.

2. Select the vertices in order and choose **Construct | Polygon Interior.**
3. Throw the dart 20 times. How many times did the dart hit the target's interior?
4. Adjust the shape of the target so the dart will always hit it. Check by throwing the dart repeatedly. Describe how you adjusted your target.
5. Adjust the shape of the target so the dart will hit it about half the time. Test your target. Describe your target and how you tested it. Include how many darts you threw and the results.
6. Make the target another shape that the dart will hit about half the time. Test your target. Describe your target and how you tested it. Include how many darts you threw and the results.
7. What determines how often the dart hits the target?
8. Select the interior of the rectangle and choose **Measure | Area.** Also measure the area of the target. What are the areas?
9. Adjust the target so that the target area is $\frac{1}{3}$ of the area of the rectangle. Throw the dart 30 times. How many times did the dart hit the target?

Dartboards

continued

CONSTRUCT

Go to page "More Dartboards." You'll see two rectangles, one on the left with a dart and an action button, and one on the right without a dart or action button. In steps 10–15, you'll construct your own virtual dart that randomly hits the rectangle on the right. In steps 16–19, you'll set up a challenge for other students.

10. On the rectangle on the right, construct points on two adjacent sides.

11. Select the two points and choose **Edit | Action Buttons | Animation.**

    In the dialog box, select one of the lines of text in the list. For Direction, choose **random,** and check the box labeled **Once only.** Now select the other line of text in the list and again choose **random** and check **Once only.** Click **OK.** Press *Animate Points* to see how the points move randomly on their segments.

12. You'll use the two random points along the sides to determine a point inside the rectangle. Select one of the points and its segment. (Make sure the *Animate Points* button is not selected.) Choose **Construct | Perpendicular Line.** Do the same with the other random point and its segment.

13. Select the two perpendicular lines you just constructed, and choose **Construct | Intersection.** Now watch the intersection point jump around as you press *Animate Points.*

14. Select the points moving randomly along the rectangle's sides, as well as the perpendicular lines, and choose **Display | Hide Objects.** Now when you press *Animate Points,* you only see the point in the rectangle's interior jump around—the tip of the dart.

15. If you want the picture of the dart on your new dartboard, copy the image in the bottom-right corner of the page. Then select the point that's to be at its tip, and choose **Edit | Paste Picture.**

16. Inside each rectangle make a target shape and its interior. Make your target a polygon or a circle.

17. Measure the areas of the interiors of the targets.

18. Select the area measurements and choose **Edit | Action Buttons | Hide/Show.** Press *Hide Area Measurements.*

19. Challenge other students to guess which target the dart is more likely to hit. (You might make it so that the chances are the same.) Allow your classmates up to 100 throws of the darts before they make their guess. Then show the area measurements so they can see if they were right.

EXPLORE MORE

20. Go to page "Last Throws." You again see two rectangles, but this time one is larger than the other. Construct points representing randomly thrown darts. Then make a target in each rectangle so the chances of hitting them are the same but don't look the same.

5

Exponents and Polynomials in Grade 8

Powering Up: Multiplication and Exponents  ACTIVITY NOTES

INTRODUCE

Project the sketch for viewing by the class. Expect to spend about 5 minutes.

1. Introduce the objectives of the activity. ***Today you'll review how to use exponents to represent repeated multiplication. Then you'll investigate the product of two powers that have the same base. I'll introduce the Sketchpad model, and then you'll work on your own.***

   Though some textbooks use the words *power* and *exponent* interchangeably, that usage can lead to confusion over "To multiply powers, add the exponents." To avoid any misunderstanding, keep the distinction clear between power and exponent as you talk with students. A *power* is a term that contains an *exponent.* For example, 10^3 is a power of 10; the exponent is 3. The expression $x + y^3$ includes a power of y. The expression $(x + y)^3$ represents the third power of $(x + y)$.

Students don't need to know how the measurement *x* is calculated from the slider for the purposes of this activity, so we refer to it simply as *x*.

2. Open **Powering Up.gsp.** Model worksheet steps 1–3. Show students how to use Sketchpad's Calculator. To create a calculation involving the x in the sketch, choose **Number | Calculate,** click on x in the sketch, use ^ to create a power of x, and then click **OK.**

3. Change the value of x by dragging the slider to test for equivalence. ***You can see that as I change the value of x, the values of the expressions $x \cdot x$ and x^2 are always equal, so they must be equivalent. Now you can try that on your own and practice using the Calculator to create powers for the rest of the repeated multiplication expressions. Then you can experiment with products.***

DEVELOP

Expect students at computers to spend 20 minutes.

4. Assign students to computers and tell them where to locate **Powering Up.gsp.**

5. Distribute the worksheet. Tell students to work through step 12 and do the Explore More if they have time. Encourage students to ask a neighbor for help if they have questions about using Sketchpad.

6. Let pairs work at their own pace. As you circulate, here are some things to notice.

   - In worksheet step 2, students may find that the sizes and styles of their calculations don't match the calculations already in the sketch. If they want their calculations to look similar to the ones in the sketch, show them how to choose **Display | Show Text Palette** and change the size and style of the calculation.

- In worksheet step 6, students don't need to create this calculation from scratch; if they try that, the Calculator will ignore the parentheses they enter. To show the parentheses, students should click on the existing expressions for $x \cdot x$ and $x \cdot x \cdot x$.
- In worksheet step 9, students should drag the slider to display the value of $x^2 \cdot x^3$ for several different values of x.

7. Encourage students to test their expressions frequently by dragging the slider. Ask questions such as, ***Can you test whether or not your expressions are equal? What do you find when you test your expressions to see whether they're equal?*** For students who are ahead of their peers, you might ask, ***Are your expressions equal even when the base, x, is negative?***

8. Some students are likely to be able to fill out the table in worksheet step 10 without testing the expressions on Sketchpad. They won't be able to verify the equivalence dynamically, so ask them to write a description explaining why their calculation process works.

SUMMARIZE

Expect to spend about 5 minutes.

9. When most students have finished worksheet step 10, direct their attention away from the computers and ask them to share the expressions they wrote in the table. Record these expressions on the board as equations, for example, $x^1 \cdot x^3 = x^4$. Ask students to suggest a product that is not in the worksheet table and ask another student to provide the simplified form. ***What is the product of $y^{20} \cdot y^4$?*** [y^{24}] ***What is $2^5 \cdot 2^4$?*** [2^9] ***What is $4^a \cdot 4^b$?*** [4^{a+b}]

10. Ask students for their rule in worksheet step 11. Work toward a complete statement and summarize: ***To multiply powers of the same base, add their exponents.*** Discuss the symbolic representation in worksheet step 12. Write it on the board.

$$x^a \cdot x^b = x^{a+b}$$

Identify this as one of several *laws of exponents* and name it according to how your students' text names it, for example, the *product of powers property*. You might ask whether the rule could be stated using other letters, and let the class come up with some examples.

11. Ask, ***How would you describe the bases in the expression in step 8 on your worksheet?*** [The expressions had the same base.] Explain that the base x could represent 2 or 3 or some other number.

12. Ask students whether they think the rule applies when bases are different. For example, ***Can $x^3 \cdot y^2$ be simplified?*** Students who had time to work on the Explore More may know that it cannot. Others may start with misconceptions, such as $x^3 \cdot y^2 = (xy)^5$. But if you ask them to defend their theories and write out the expressions as repeated multiplication, they will probably conclude that the law applies only to like bases and that $x^3 \cdot y^2$ cannot be simplified.

13. If time permits, give students some guided practice finding products of powers, perhaps including some monomials with numerical coefficients and/or more than one base.

ANSWERS

5. $x \cdot x = x^2$; $x \cdot x \cdot x = x^3$; $x \cdot x \cdot x \cdot x = x^4$; $x \cdot x \cdot x \cdot x \cdot x = x^5$; $x \cdot x \cdot x \cdot x \cdot x \cdot x = x^6$; $x \cdot x \cdot x \cdot x \cdot x \cdot x \cdot x = x^7$

7. $(x \cdot x) \cdot (x \cdot x \cdot x) = x^2 \cdot x^3$

8. $(x \cdot x)\ (x \cdot x \cdot x) = x^2 \cdot x^3 = x^5$

10.

Product	Product of Powers $x^? \cdot x^?$	Simplified Power $x^?$
$(x) \cdot (x \cdot x \cdot x)$	$x^1 \cdot x^3$	x^4
$(x \cdot x) \cdot (x \cdot x \cdot x \cdot x)$	$x^2 \cdot x^4$	x^6
$(x \cdot x) \cdot (x \cdot x)$	$x^2 \cdot x^2$	x^4
$(x \cdot x \cdot x \cdot x) \cdot (x \cdot x \cdot x)$	$x^4 \cdot x^3$	x^7
$(x \cdot x \cdot x \cdot x) \cdot (x \cdot x \cdot x \cdot x)$	$x^4 \cdot x^4$	x^8
$(x \cdot x) \cdot (x \cdot x) \cdot (x \cdot x \cdot x)$	$x^2 \cdot x^2 \cdot x^3$	x^7

11. When you multiply powers with like bases, you add exponents.

12. $x^a \cdot x^b = x^{a+b}$

13. Students who try the Explore More should discover that products of powers with different bases can't be multiplied by adding the exponents.

Powering Up

Name:

In this activity you'll experiment with multiplying powers with the same base. You'll discover a rule for writing products of powers.

EXPLORE

1. Open **Powering Up.gsp** and go to page "Multiplying Powers." You'll see the equation $x = 2$ and equations containing two factors of x, three factors of x, and other products of repeated multiplication.

2. One factor of x can be written with an exponent, x^1, as shown in the sketch. Choose **Number | Calculate.** In the Calculator, create an expression that uses an exponent and is equivalent to $x \cdot x$.

   To enter x into the Calculator, click on the equation $x = 2$ in the sketch. To raise it to a power, click ^ on the Calculator keypad.

   When you're finished, click **OK.** If necessary, drag the calculation to make it line up nicely under $x^1 = 2$.

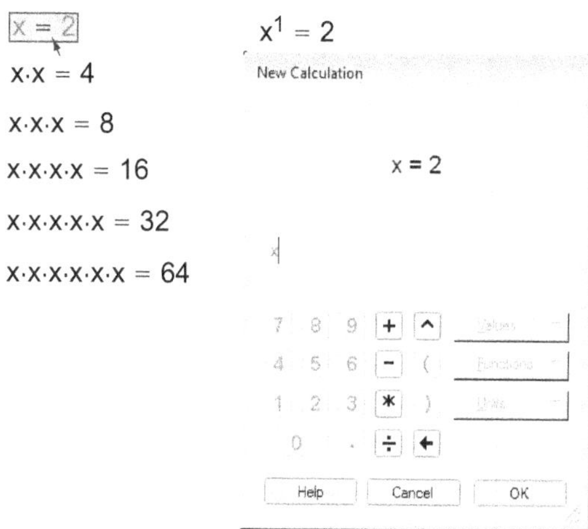

3. To check whether the new expression is equivalent to $x \cdot x$, change the value of x by dragging the slider, and look at the values of both expressions.

4. Is the expression you created equivalent to $x \cdot x$?

5. Repeat step 2 to create expressions with exponents that are equivalent to each of the other products of repeated multiplication. Check for equivalence by changing the value of x.

6. Now you'll experiment with products of powers. Choose **Number | Calculate** and click on $x \cdot x$, then $*$, and then $x \cdot x \cdot x$ to create the expression $(x \cdot x)(x \cdot x \cdot x)$. Click **OK.**

7. To create an expression that is equal to the expression in step 6, choose **Number | Calculate** and enter the product of two powers of x.

   Write this equation with the correct exponents.

$$(x \cdot x) \cdot (x \cdot x \cdot x) = x^? \cdot x^?$$

8. What expression with a single exponent is the product in step 6 equal to? Write this equation with the correct exponents.

$$(x \cdot x) \cdot (x \cdot x \cdot x) = x^? \cdot x^? = x^?$$

9. Check that your expressions in step 8 are equal by changing the value of x.

10. Repeat steps 6–9 for each of the products in the table. Drag the slider to test that both expressions are always equal. Record your expressions in the table.

Product	Product of Powers $x^? \cdot x^?$	Simplified Power $x^?$
$(x) \cdot (x \cdot x \cdot x)$		
$(x \cdot x) \cdot (x \cdot x \cdot x \cdot x)$		
$(x \cdot x) \cdot (x \cdot x)$		
$(x \cdot x \cdot x \cdot x) \cdot (x \cdot x \cdot x)$		
$(x \cdot x \cdot x \cdot x) \cdot (x \cdot x \cdot x \cdot x)$		
$(x \cdot x) \cdot (x \cdot x) \cdot (x \cdot x \cdot x)$		

11. Look for patterns in your table. Describe how to multiply powers with the same base.

12. Replace the question mark to generalize your observation as a rule.

$$x^a \cdot x^b = ?$$

EXPLORE MORE

13. Choose **Number | New Parameter** and name the new variable *y*. Make its value different from *x*. Calculate a power of *y*.

    Experiment with products of powers of *x* and *y*. Is it possible to simplify a product of two powers such as $x^2 \cdot y^3$? Can you create an equivalent expression with just one exponent? Explain.

Powering Down: Division and Exponents

ACTIVITY NOTES

INTRODUCE

Project the sketch for viewing by the class. Expect to spend about 5 minutes.

1. Students should have completed the activity Powering Up. Explain, ***Today you'll review a simple but important rule about dividing powers. Then you'll experiment with dividing two powers of the same base. I'll show you how to divide using Sketchpad's Calculator, and then you'll work on your own.***

   Though some textbooks use the words *exponent* and *power* interchangeably, that usage can lead to confusion over "To divide powers, subtract the exponents." Make the distinction in your own use of the words: A *power* is a term that contains an *exponent.* For example, 10^3 is a power of 10, and the exponent is 3. The expression $x + y^3$ includes a power of y. The expression $(x + y)^3$ represents the third power of $(x + y)$.

Students don't need to know how the measurement x is calculated from the slider for the purposes of this activity, so we refer to it simply as x.

2. Open **Powering Down.gsp.** Model worksheet steps 1–3. Show students how to use Sketchpad's Calculator. To create a calculation involving the x in the sketch, such as $\frac{x}{x}$, choose **Number | Calculate,** click on x in the sketch, then ÷ on the Calculator keypad, then x again, and then click **OK.** Pressing *Show Powers* displays the expressions using exponents.

3. ***What do I get if I divide x by x? Does it matter what the value of x is?*** Demonstrate how to change the value of x by dragging the slider. ***What number should I substitute for x?*** Try several values students suggest. If no student suggests 0, you might suggest it to show that $\frac{0}{0}$ is undefined. You might ask, ***Why is $\frac{0}{0}$ undefined?*** [Because any number times 0 is 0, there is no way to determine the answer: $0 \cdot 5 = 0$; $0 \cdot 154 = 0$; $0 \cdot n = 0$, so $\frac{0}{0}$ is 5 or 154 or any number.]

4. ***As long as x is not 0, when I divide it by itself, I get 1. Can you think of an example in arithmetic where you use this idea?*** If necessary, follow up with, ***How do we use this idea when we reduce fractions?***

5. ***You will now use the idea of dividing out common factors as you simplify expressions that involve quotients of powers.***

DEVELOP

Expect students at computers to spend 20 minutes.

6. Assign students to computers and tell them where to locate **Powering Down.gsp.**

7. Distribute the worksheet. Tell students to work through step 12 of the worksheet and do the Explore More if they have time. Encourage

students to ask a neighbor for help if they have questions about using Sketchpad.

8. Let pairs work at their own pace. As you circulate, here are some things to notice.
   - As students work on worksheet step 4, ask, ***When have you used this law of arithmetic?*** [Reducing fractions]
   - In worksheet step 5, students can create the calculation from scratch or they can use the existing expressions, for $x \cdot x \cdot x \cdot x \cdot x$ and $x \cdot x \cdot x$. If they use the existing expressions, the numerator and denominator will appear with parentheses; if they create a calculation from scratch, there will be no parentheses.
   - In worksheet step 8, students should calculate $\frac{x^5}{x^3}$ and compare it to x^2.
   - For the expression $\frac{x \cdot x \cdot x \cdot x \cdot x}{x}$ in worksheet step 9, ask, ***How can the denominator, x, be written with an exponent?*** [x^1]
9. Encourage students to test their expressions by changing the value of x. ***What do you find when you test your expressions to see whether they're equivalent?***
10. Some students may be able to fill out the table in worksheet step 9 without testing the expressions on Sketchpad. Ask them to explain why their methods work. To add more challenges, you might ask, ***Are your expressions equivalent even when the base, x, is negative?***

SUMMARIZE

Expect to spend 5 minutes.

11. When students have finished worksheet step 10, direct their attention away from the computers and write $\frac{x \cdot x \cdot x \cdot x \cdot x}{x \cdot x \cdot x}$ on the board. Ask them to explain why $\frac{x^5}{x^3} = x^2$. Be sure the discussion includes that $\frac{x}{x}$ equals 1, so each pair of factors of x in the numerator and denominator divide to make 1. (That is, the x's "cancel.") For example, in worksheet step 7, two factors of x remain in the numerator.
12. Have students share the expressions they wrote in the table in worksheet step 9. Record these expressions on the board as equations: $\frac{x^5}{x^2} = x^3$, for example.

13. Ask what students gave for their rule in worksheet step 11. Write it on the board.

$$\frac{x^a}{x^b} = x^{a-b}$$

Identify this as one of several *laws of exponents* and name it according to its name in your students' text, for example, the *quotient of powers property.* Clearly state the rule in words. ***To divide powers of the same base, subtract the exponents.*** Include the phrase *of the same base.*

14. Point out that in the investigation, all the expressions had the same base, x. Ask students, ***Do you think the rule applies when bases are different? For example, can you simplify*** $\frac{x^3}{y^2}$***?*** Students may start with misconceptions, but if you ask them to defend their theories, they should conclude that no common factors will divide out, and the law only applies to like bases.

15. If time permits, give students some guided practice dividing powers, perhaps including monomials with numerical coefficients or expressions with more than one base.

16. If negative exponents are part of your curriculum, discuss Explore More.

ANSWERS

4. Any number divided by itself is equal to 1.

6. $\frac{x \cdot x \cdot x \cdot x \cdot x}{x \cdot x \cdot x} = x \cdot x = x^2$

7. The three factors of x in the numerator divided by three factors of x in the denominator equal 1. Two factors of x remain in the numerator, so the quotient equals $x \cdot x$.

8. $\frac{x \cdot x \cdot x \cdot x \cdot x}{x \cdot x \cdot x} = \frac{x^5}{x^3} = x^2$

9.

Quotient Using Repeated Factors	Equivalent Quotient Using Exponents $\frac{x^?}{x^?}$	Equivalent Power $x^?$
$\frac{x \cdot x \cdot x \cdot x \cdot x}{x \cdot x}$	$\frac{x^5}{x^2}$	x^3
$\frac{x \cdot x \cdot x \cdot x}{x \cdot x \cdot x}$	$\frac{x^4}{x^3}$	x^1 or x
$\frac{x \cdot x \cdot x \cdot x \cdot x}{x}$	$\frac{x^5}{x^1}$	x^4
$\frac{x \cdot x \cdot x \cdot x \cdot x \cdot x}{x \cdot x}$	$\frac{x^6}{x^2}$	x^4
$\frac{x \cdot x \cdot x}{x \cdot x \cdot x}$	$\frac{x^3}{x^3}$	x^0 or 1

10. The exponent in the equivalent power is the difference between the exponent in the numerator and the exponent in the denominator.

11. $\frac{x^a}{x^b} = x^{a-b}$

12. $x^0 = 1$

13. They should conclude that $x^{-n} = \frac{1}{x^n}$.

Powering Down

Name:

In this activity you'll experiment with dividing powers with the same base. You'll discover a rule for writing quotients of powers.

EXPLORE

1. Open **Powering Down.gsp** and go to page "Dividing Powers." You'll see the equation $x = 4$ and equations containing two factors of x, three factors of x, and other products of repeated multiplication. Each repeated multiplication of x can be written as a power of x.

2. Now you will create the calculation $\frac{x}{x}$. Choose **Number | Calculate.** In the Calculator, enter x by clicking on the equation $x = 4$ in the sketch. Then click ÷ on the Calculator keypad, and then click on x again.

   When you're finished, click **OK.**

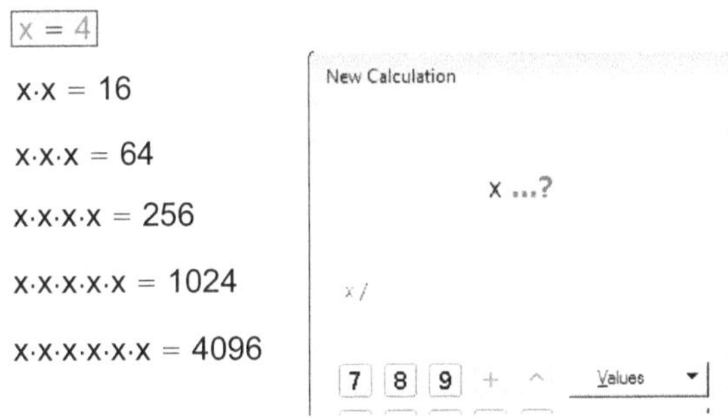

3. Change the value of x by dragging the slider.

4. What basic law of arithmetic does this calculation demonstrate?

5. Now, you'll experiment with dividing powers. Choose **Number | Calculate** and click on $x \cdot x \cdot x \cdot x \cdot x$ in the sketch, ÷ on the keypad, $x \cdot x \cdot x$ in the sketch, and then **OK** to create the expression

$$\frac{x \cdot x \cdot x \cdot x \cdot x}{x \cdot x \cdot x}$$

6. The calculation should be equal to one of the calculations already in your sketch. Which one? Drag the slider to confirm that these expressions are always equal.

7. How does your answer to step 4 help explain why $\frac{x \cdot x \cdot x \cdot x \cdot x}{x \cdot x \cdot x} = x \cdot x$?

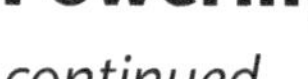

8. Choose **Number | Calculate.** Create another expression equivalent to $\frac{x \cdot x \cdot x \cdot x \cdot x}{x \cdot x \cdot x}$ in the form of $\frac{x^?}{x^?}$. Use the x in the sketch and the ^ on the Calculator keypad to enter powers. As you finish each expression, click **OK.**

   Write this equation with the correct exponents.

$$\frac{x \cdot x \cdot x \cdot x \cdot x}{x \cdot x \cdot x} = \frac{x^?}{x^?} = x^?$$

9. Repeat steps 5 and 8 for each of the quotients in the table. Drag the slider to test that both expressions are always equal. Record your expressions in the table.

Quotient Using Repeated Factors	Equivalent Quotient Using Exponents $\frac{x^?}{x^?}$	Equivalent Power $x^?$
$\frac{x \cdot x \cdot x \cdot x \cdot x}{x \cdot x}$		
$\frac{x \cdot x \cdot x \cdot x}{x \cdot x \cdot x}$		
$\frac{x \cdot x \cdot x \cdot x \cdot x}{x}$		
$\frac{x \cdot x \cdot x \cdot x \cdot x \cdot x}{x \cdot x}$		
$\frac{x \cdot x \cdot x}{x \cdot x \cdot x}$		

10. Look for patterns in your table. Describe how to divide powers with the same base.

11. Replace the question mark to generalize your observation as a rule.

$$\frac{x^a}{x^b} = ?$$

12. The last row of your table demonstrates a special case of the rule that you might find surprising. Complete this equation.

$$x^0 = \_\_$$

EXPLORE MORE

13. In all the quotients in step 9, the power in the numerator has a larger exponent than the power in the denominator. What if the power in the denominator had the larger exponent? Experiment calculating expressions like $\frac{x^3}{x^6}$. Is the generalization you made in step 11 still true? Experiment calculating powers with negative exponents. Complete this equation.

$$x^{-n} = \_\_$$

INTRODUCE

Project the sketch for viewing by the class. Expect to spend about 10 minutes.

1. Open **Power Strips.gsp** and go to page "Patterns." Enlarge the document window so it fills most of the screen.
2. Explain, ***Today you're going to use Sketchpad to explore operations involving exponents. You'll look for patterns and make conjectures.***
3. Slowly drag the red point x left and right on the sketch. ***What do you observe? What do you think the points represent?*** Let students share their ideas; don't give specific feedback at this time. These questions encourage students to observe closely and think about the patterns they see.
4. Press *Show Expressions.* Help students recognize that the points represent various powers of x. The vertical position corresponds to the power (exponent) and the horizontal position corresponds to the value of x. Tell students they will use this model to explore operations with exponents.
5. Continue dragging the red point x. ***How many different positions make all the points line up? What values of x do you think these positions indicate?*** [The points all line up when $x = 0$ and $x = 1$.] Let students conjecture, then press *Show Numbers.* Have students explain why the points should line up at these particular values.
6. ***These expressions can be abbreviated using exponents. Can you explain what an exponent means?*** Work with the class to come up with a definition of *exponent* and write it on the board. Here is a sample definition: *An exponent is a mathematical notation that means a number is multiplied by itself repeatedly. The base is the number to be multiplied, and the exponent is how many times the base is a factor. The number expressed as a base raised to an exponent is called a power.* Show an example, such as 5^3. Make sure students understand that it represents $5 \cdot 5 \cdot 5$, that 5 is the base, and that 3 is the exponent. The expression 5^3 can be read "the third power of 5" or "5 to the third power." The latter uses the word *power* to mean *exponent,* which may be confusing to students.
7. Go to page "Multiplication." Model worksheet step 9, ***Now I'll show you how to use one of the custom tools, the* a*b *tool. After you choose this tool, you must click on five objects in this order: (1) an unused point, (2) the point of the first strip, (3) the label of this point, (4) the point of***

the second strip, and (5) the label of this point. Notice that the tool name includes these five steps. Emphasize that an "unused point" is one of the small circles that is not already the left endpoint of a strip. Make sure you have practiced using the tool beforehand. The other custom tools work in the same way.

8. If you want students to save their work, demonstrate choosing **File | Save As,** and let them know how to name and where to save their files.

DEVELOP

Expect students at computers to spend about 25 minutes.

9. Assign students to computers and tell them where to locate **Power Strips.gsp.** Distribute the worksheet. Tell students to work through step 22 and do the Explore More if they have time. Encourage students to ask their neighbors for help if they are having difficulty with Sketchpad.

10. Let pairs work at their own pace. As you circulate, here are some things to notice.

    - In worksheet steps 5 and 6, be sure students notice that the differences in the lengths of the strips are not constant. As x moves from 1 to the right, the lengths of the strips are increasing. As x moves from 1 to 0, the lengths of the strips decrease. ***Are the differences in the lengths of the strips constant from one strip to the next?***

    - In worksheet step 7, students should notice that the strips alternate, pointing right and then left. ***What is the sign of the product when you multiply two negative numbers? When you multiply three negative numbers?***

    - Before students go to worksheet step 8, have students tell you how to write the expression $x \cdot x \cdot x \cdot x$ using exponents. [x^4] This will help them connect page "Multiplication" with page "Patterns." The model is the same, but the "Multiplication" page uses exponents.

    - In worksheet steps 9–22, students will use custom tools to multiply and divide powers with the same base and find the value of a power raised to a power. Have students help each other use the tools if questions arise.

    - In worksheet step 12, encourage students to drag x back and forth to check that the x^5 strip is the only strip that is the same length. ***When you test, be sure to drag x left and right to try different values.***

Ask students to write out x^3, x^2, and x^5 using repeated multiplication to see why the strip for x^5 matches.

- In worksheet steps 16 and 20, have students use repeated multiplication to see why these problems work as well. ***Can you rewrite these expressions without using exponents?***
- In worksheet steps 14, 18, and 22, students will try to make conjectures after observing the simplification of two sample expressions. Encourage students to construct as many expressions as needed to make and test their conjectures. ***If you don't see a pattern yet, make another expression. Continue until you can make a conjecture. Then test it.***
- If students have time for the Explore More, they will investigate what happens when multiplying powers with different bases. Students will discover that there is no general pattern, so no rule can be made about simplifying a problem like $x^a \cdot y^b$.

11. If students will save their work, remind them where to save it now.

SUMMARIZE

Project the sketch. Expect to spend about 10 minutes.

12. Gather the class. Open **Power Strips.gsp.** Students should have their worksheets with them. Begin the discussion by asking students to share their conjectures. ***What conjectures did you make today?*** Volunteers may wish to test their conjectures for the class by constructing a problem using the appropriate page on the sketch.

13. When the class agrees on a conjecture, write it on the board using words and symbols. You may need to help students observe that the bases were the same in the problems. ***What do you notice about the bases in all the problems?*** Here is a sample summary of the properties with exponents covered in this activity.

Words	Symbols
When multiplying two powers with the same base, keep the base and add the exponents.	$x^a \cdot x^b = x^{a+b}$
When dividing two powers with the same base, keep the base and subtract the exponents.	$\frac{x^a}{x^b} = x^{a-b}$
When raising a power to a power, keep the base and multiply the exponents.	$(x^a)^b = x^{ab}$

14. If time permits, discuss the Explore More. ***What did you notice when the bases were different? Could you still use your conjecture from worksheet step 14?*** It is important that students understand that the bases need to be the same for properties to hold true. There is no rule for simplifying problems with different bases.

15. You may wish to have students respond individually in writing to this prompt. ***Are $(x^a)^b$ and $x^a \cdot x^b$ the same value? Explain.*** [No, $(x^a)^b = x^{ab}$ and $x^a \cdot x^b = x^{a+b}$.]

EXTEND

What questions occurred to you about operations with exponents? Encourage curiosity. Here are some sample student queries.

Why do the rules hold?

What happens when the exponents are negative? Do the properties still work?

Is $(x^a)^b$ the same as $(x^b)^a$?

Can exponents ever be fractional, between integers?

When an exponent is 0, the expression simplifies to 1. Does this affect the properties at all?

What happens when you have a more complicated expression like $(x^2y^3)^2$?

How do you simplify an expression like $\left(\frac{3x^2}{y^2}\right)^3$?

ANSWERS

3. Answers will vary. The points represent various powers of x, with the vertical position corresponding to the power and the horizontal position corresponding to the value of x.

4. There are two positions of x at which all the points line up: $x = 1$ and $x = 0$. Repeatedly multiplying either 1 or 0 by itself continues to give the same result.

5. When $x > 1$, the points move increasingly rightward as you go down the screen, showing that the value of x^n increases more and more quickly for greater and greater values of n.

6. When $0 < x < 1$, the points move toward the left as they go down, approaching a straight line at $x = 0$. This pattern makes sense because multiplying by a value less than 1 always gives a result that is closer to 0 than the number you started with.

7. When $x < 0$, the strips alternate between the left and right sides because multiplying a number by a negative value always gives a result with a sign opposite to the sign of the original number.

10. The $x^3 \cdot x^2$ strip is the same length as the x^5 strip.

12. The $x^3 \cdot x^2$ strip is the same length as the x^5 strip no matter how you drag the value of x. This makes sense because you've multiplied three x's by two more x's, so that there are now five x's multiplied together.

13. The strip for $x^4 \cdot x^3$ is the same length as the x^7 strip.

14. Conjecture: The strip for $x^a \cdot x^b$ is the same length as the x^{a+b} strip. When multiplying powers with the same base, keep the base and add the exponents. Students will construct different products to test this conjecture.

16. When you use the **a/b** tool to create a strip for $\frac{x^7}{x^3}$, the resulting strip is the same length as the x^4 strip.

17. The resulting strip is the same length as the x^6 strip.

18. Conjecture: The strip for $\frac{x^a}{x^b}$ is the same length as the x^{a-b} strip. When dividing powers with the same base, keep the base and subtract the exponents. Students will construct different quotients to test this conjecture.

20. When you use the **(a)^b** tool to create a strip for $(x^4)^2$, the result matches x^8.

21. The resulting strip is the same length as the x^9 strip.

22. Conjecture: The strip for $(x^a)^b$ is the same length as an x^{ab} strip. When raising a power to a power, keep the base and multiply the exponents. Students will construct different powers of powers to test this conjecture.

23. If the bases are not the same, there is no general rule you can use to simplify a problem like $x^a \cdot y^b$. The special case of $x^a \cdot y^a = (xy)^a$ does not generalize to cases in which $a \neq b$, as testing with many values of x, y, a, and b will confirm.

Power Strips

Name:

In this activity you'll use a visual model to explore operations involving exponents.

EXPLORE

1. Open **Power Strips.gsp.** Go to page "Patterns."

2. Drag the red point x left and right. Observe what happens to the green points.

3. Press *Show Expressions.* Explain what the points represent.

4. How many positions can you find that make all the points line up? What values of x do you think these positions indicate? Press *Show Numbers.* Explain why the points line up for these values.

x
$x \cdot x$
$x \cdot x \cdot x$
$x \cdot x \cdot x \cdot x$
$x \cdot x \cdot x \cdot x \cdot x$
$x \cdot x \cdot x \cdot x \cdot x \cdot x$
$x \cdot x \cdot x \cdot x \cdot x \cdot x \cdot x$
$x \cdot x \cdot x \cdot x \cdot x \cdot x \cdot x \cdot x$
$x \cdot x \cdot x \cdot x \cdot x \cdot x \cdot x \cdot x \cdot x$

5. What pattern do the points make when $x > 1$? Why does this make sense?

6. What pattern appears when $0 < x < 1$? Explain this pattern.

7. What pattern appears when $x < 0$? Explain this pattern.

8. Go to page "Multiplication." This page uses exponents to write the multiplication problems more simply. Drag point x. You should observe the same behaviors as on page "Patterns."

9. Choose the **a*b** custom tool from the Custom Tools menu. Use the **a*b** tool to multiply $x^3 \cdot x^2$ by clicking on these five points in order:

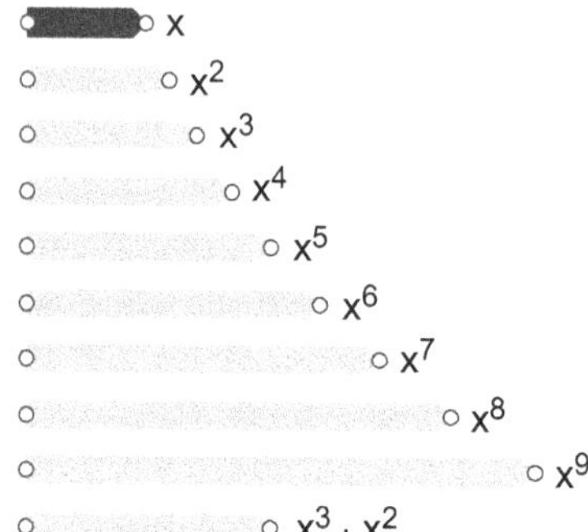

- The first unused point below all the strips
- The point at the tip of the x^3 strip
- The x^3 label
- The point at the tip of the x^2 strip
- The x^2 label

10. Is the $x^3 \cdot x^2$ strip the same length as any of the existing strips?

11. Test your answer by clicking the **Indicator** custom tool on the tip of the $x^3 \cdot x^2$ strip.

12. Drag x and describe your observations.

13. Multiply $x^4 \cdot x^3$. What existing strip does it match for all values of x?

14. Make a conjecture for $x^a \cdot x^b$. Test your conjecture by constructing another product of two powers.

15. Now go to page "Division." You'll use this page to explore division problems.

16. Use the **a/b** custom tool to construct a strip for $\frac{x^7}{x^3}$. What is the result?

17. Now find $\frac{x^8}{x^2}$. What is the result?

18. Make a conjecture for $\frac{x^a}{x^b}$. Test your conjecture by constructing another quotient of powers.

19. Go to page "Powers." You'll use this page to explore problems like $(x^4)^2$.

20. Use the **(a)^b** custom tool to construct a strip for $(x^4)^2$. What is the result?

21. Now find $(x^3)^3$. What is the result?

22. Make a conjecture for $(x^a)^b$. Test your conjecture by constructing another power of powers.

EXPLORE MORE

23. Go to page "Explore More." What if the bases are not the same? Is there a rule for problems like $x^a \cdot y^b$? Experiment and then describe your conclusions.

Broccoli and Brussels Sprouts: The Distributive Property

ACTIVITY NOTES

INTRODUCE

Project the sketch for viewing by the class. Expect to spend about 10 minutes.

1. Open **Broccoli & Brussels Sprouts.gsp** and go to page "Diagram." Enlarge the sketch to fill the screen and say, ***This diagram shows a garden that is divided into two parts, one for broccoli and one for Brussels sprouts. Each centimeter in the diagram represents one meter in the real garden. The head gardener and her assistant needed to find the total area of the garden. They calculated it in two different ways, but both got the same answer. What do you think the two ways were?*** A student might answer, *One added 15 and 10 and then multiplied by 14, and the other multiplied 14 times 15 and 14 times 10 and then added.* Have students share, either verbally or by writing on the board, how the calculation methods give the same answer.

$$14 \cdot (15 + 10) = 14 \cdot 25 = 350$$

$$14 \cdot 15 + 14 \cdot 10 = 210 + 140 = 350$$

Expect students to write or say these calculations in different ways, but summarize by writing the equivalent calculations as an equation.

$$14 \cdot (15 + 10) = 14 \cdot 15 + 14 \cdot 10$$

2. Explain, ***This is called the* distributive property. *Today you are going to use dynamic algebra tiles to model situations like these. Notice that the garden values can change.*** Drag the red points to show that the measured values update to reflect the actual distances.

3. ***On this page, you're going to make two calculations that both give an answer equal to the total area.*** Demonstrate how to enter a measured value in a calculation. Choose **Number | Calculate** and click on a value in the sketch to enter it in the Calculator, but don't complete the expression—leave that for the students.

4. Go to page "Tiles." Explain, ***In the rest of the activity, you'll be using dynamic algebra tiles to model distributive property expressions. On this page, the expressions are already modeled. The tiles on the outside of the corner piece represent the dimensions of the rectangle. The expression $3(x + 2)$ can be thought of as length multiplied by width. Your job will be to build and write the missing expression representing the area of the rectangle inside the frame.***

5. Go to page "Practice" and say, ***On the rest of the pages, you'll use custom tools to construct the models yourself.*** Demonstrate how to create the

arrangement shown on page "Tiles." You should also demonstrate the following mistake in some blank space on the page: ***A common mistake is to try to line up unit tiles with the variable edge of an x tile. But notice that if x changes, the units don't line up anymore.***

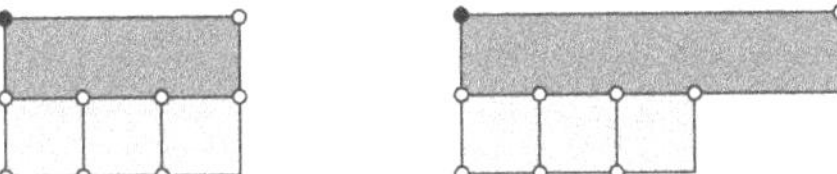

You can line up unit tiles with the unit edges of the x tiles.

6. Go to page "GCF" and say, ***On this page you're given the rectangle and the area expression. You'll put tiles along the outside of the frame piece to represent the dimensions and write the expression length multiplied by width. On the rest of the pages, you'll be building the rectangles inside the frame and the dimensions along the outside of the frame.***
7. If you want students to save their work, demonstrate choosing **File | Save As,** and let them know how to name and where to save their files.

DEVELOP

Expect students at computers to spend about 30 minutes.

8. Distribute the worksheet and assign students to computers. Tell them where to locate **Broccoli & Brussels Sprouts.gsp.** Tell students to work through step 9 and do the Explore More if they have time. Encourage students to ask their neighbors for help if they are having difficulty with the construction.
9. Let pairs work at their own pace. As you circulate, here are some things to notice.
   - In worksheet steps 2 and 5, look for students who may need help entering an expression in the Calculator. A common mistake is to enter numbers or expressions instead of clicking on them in the sketch.
   - Watch for students who try to line up unit tiles with the variable dimension of an *x* tile. (See step 5 above.)
   - Remind students they can always undo if they make a mistake or accidentally create a stray tile.

- In parts of worksheet step 9, students will need to use the **x^2** tool to construct the rectangles in the frame.
- In worksheet steps 8 and 9, remind students to test their expressions for equivalence by calculating the two expressions on each page.

10. If students will save their work, remind them where to save it now.

SUMMARIZE

Project the sketch. Expect to spend about 5 minutes.

11. Gather the class. Display the sketch **Broccoli & Brussels Sprouts.gsp** again and go to page "Garden." Ask students what equal expressions they calculated. Write the equation $a(b + c) = ab + ac$, and identify this as the distributive property of multiplication over addition.
12. If a student did a Make Your Own problem, invite her to share it on the overhead display. Have her construct the model without revealing the expressions it represents; then invite the class to come up with the equation.
13. If your curriculum uses the vocabulary *factored form* and *expanded form*, this is a good opportunity to introduce it. Use a student example or an example from the worksheet. Use worksheet step 9e as an example. ***In the equation $x^2 + 7x = x(x + 7)$, the left side expression is called*** **expanded form** ***and the right side is called*** **factored form.** ***The factors are x and x + 7. In steps 9a through 9c, in what form were the given expressions?*** (Factored form) ***What about in steps 9d through 9f?*** (Expanded form)

ANSWERS

3. $a*(b + c) = a*b + a*c$
4. $3(x + 2) = 3x + 6$
8. $4(x + 1) = 4x + 4$
9. a. $3(2x + 1) = 6x + 3$
   b. $3x(x + 2) = 3x^2 + 6x$
   c. $x(2x + 5) = 2x^2 + 5x$
   d. $2x + 8= 2(x + 4)$
   e. $x^2 + 7x = x(x + 7)$
   f. $3x^2 + 3x = 3x(x + 1)$

10. Possible expressions that can be modeled with rectangles are $3x(x + 1)$ and $x(3x + 3)$. Students might also observe that $3(x^2 + x)$ is another equivalent expression.
11. $a(b + c) = ab + ac$

Broccoli and Brussels Sprouts

Name:

The garden at right is divided into two parts, one for broccoli and one for Brussels sprouts. The head gardener and her assistant needed to know the total area of the garden. They each calculated it and got the same answer, but they did their calculations in two different ways.

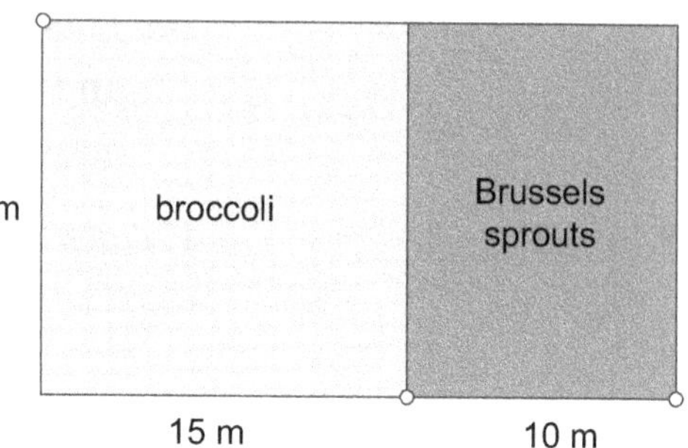

In this activity you'll model the distributive property—first with a dynamic garden, and then using dynamic algebra tiles.

EXPLORE

1. Open **Broccoli & Brussels Sprouts.gsp** and go to page "Diagram."

2. Choose **Number | Calculate** and enter an expression that is equal to the total area. Click on the measurements in the sketch to enter them into the Calculator.

3. Use the Calculator to enter a different expression that is equal to the total area. Drag points to confirm that both expressions are always equal to the total area. Write the equivalent expressions as an equation.

   ____________ = ____________

4. Go to page "Tiles." The tiles on the outside of the frame represent the dimensions of the rectangle inside the frame. Complete the equation by writing the expression represented by the tiles inside the frame.

   $3(x + 2) =$ ____________

5. Choose **Number | Calculate** and enter the expression $3*(x + 2)$ into the Calculator. Click on the x-value on the slider to enter it in the calculation. Click **OK.**

6. Now enter the expression you wrote in step 4. Drag the x slider. Do the two expressions always remain equal?

7. Go to page "Practice." Practice using the custom tools to construct the arrangement on page "Tiles."

Broccoli and Brussels Sprouts

continued

8. Go to page "GCF." The tiles inside the frame are represented by the expression $4x + 4$. Attach tiles along the outside of the frame to represent the dimensions of the rectangle. Complete the equation.

$$\_\_\_\_\_\_\_\_\_\_ = 4x + 4$$

9. On pages "a" through "f," follow these steps.

   - Use tiles to construct the dimensions along the outside of the frame and the rectangle inside the frame.
   - Write the equation using the two representations of the area.
   - Use Sketchpad's Calculator to calculate both expressions.
   - Drag the *x* slider to check that the expression you wrote is always equivalent to the given expression.

   a. $3(2x + 1) =$ __________

   b. $3x(x + 2) =$ __________

   c. $x(2x + 5) =$ __________

   d. $2x + 8 =$ __________

   e. $x^2 + 7x =$ __________

   f. $3x^2 + 3x =$ __________

EXPLORE MORE

10. For the problem on page "f," there are two different-shaped rectangles you can make. Make the other possible rectangle and write the equation for this rectangle.

$$3x^2 + 3x = \_\_\_\_\_\_\_\_\_\_$$

11. Go to page "Garden" and represent the garden problem as an equation using only variables.

12. Go to page "Make Your Own." Use the tiles to create models of your own distributive property equation.

This activity can be used by students whether or not they have had hands-on experience with physical algebra tiles. Most students, especially tactile learners, benefit from manipulating physical tiles. This activity offers a similar model, using Sketchpad tiles to represent polynomials.

If your students have experience with physical tiles, you might have them use custom tools to construct arrangements that will update dynamically as students drag sliders for x and y. If so, students should use page "Explore More," rather than page "Frame," to complete the worksheet. Students who do not have experience with physical tiles will need to work with the Sketchpad tiles before they can effectively use the custom tools.

INTRODUCE

Project the sketch for viewing by the class. Expect to spend about 10 minutes.

1. The first page of the student worksheet—"Get Ready"—can be done in class the day before students use computers, assigned as homework, or used as a warm-up. If your students have never used algebra tiles, you should introduce them, as explained in the next step, *before* assigning Get Ready. In any case, you should review the Get Ready questions before starting the activity. Ask for volunteers to explain their answers. Emphasize the relationship between dimensions and area.
2. Open **Tiling in a Frame.gsp** and go to page "Tiles." Make sure that the dimensions are showing. Drag tiles next to other tiles to compare their dimensions. Explain that tiles are named after their areas and make sure students can identify all the names. Drag the x and y sliders to show that the area expressions match the dimensions for all values. Explain that this activity uses rigid tiles, but that students can explore dynamic tiles in the Explore More.
3. Go to page "Example 1" (or go to page "Frame" and make up your own examples). Enlarge the document window so it fills most of the screen. Explain, ***Today you're going to use Sketchpad to create rectangles using tiles. First you'll drag tiles from the stacks to the outside edges of the frame. Here you can see $x + 2$ along the top and $x + 3$ along the side. These are the dimensions of the rectangle.*** Students should pay attention to how these *dimension tiles* are represented on the frame—along the top and outside left edges of the frame, and separated by like terms.

4. Explain, ***You'll drag tiles from the stacks into the frame to create rectangles. The tiles should be arranged so that there are no gaps or overlaps.*** Press *Show Tile Area.* Explain that the dimensions of this rectangle match those along the outside of the frame. The "seams" in the completed rectangle create straight segments that extend the entire length or width of the rectangle and are aligned with the "seams" in the dimension tiles. Press *Show Blueprint Lines* if you want to emphasize this, but students will not have blueprint lines for the problems on the worksheet. You can show the blueprint lines and tile areas simultaneously to demonstrate that rays constructed from the dimension tiles overlap every seam without going through the interior of any of the tiles.

5. Ask, ***What is the area of the completed rectangle?*** Then use a text caption, or use the board, to write the multiplication problem.

$$length \times width = area$$

$$(\text{or, } base \times height = area)$$

$$(x + 2)(x + 3) = x^2 + 5x + 6$$

Ask what each polynomial represents. Make sure that students know their basic multiplication terminology and notation. They should know what *factors* are and that a product can also be symbolized using only parentheses (without the $\times$ symbol).

6. If you wish to give more guidance, go to page "Example 2." You might review the example together with the class, ask student pairs to review it on their own, or mention it to student pairs only if they need support.

DEVELOP

Expect students at computers to spend about 30 minutes.

7. Assign students to computers and tell them where to locate **Tiling in a Frame.gsp.** Distribute the worksheet. Tell students to work through step 14 and do the Explore More if they have time. Encourage students to ask their neighbors for help if they are having difficulty with Sketchpad.

8. Let pairs work at their own pace. As you circulate, here are some things to notice.

   - Make sure students are placing tiles so that they match both dimensions along the outside of the frame. In particular, make sure students are not using 3 unit tiles as the equivalent of an *x* tile.

- If a student pair is really struggling, you might show them how to construct blueprint lines by using the **Ray** tool to click on the intersections of the adjacent tiles along the outside of the frame.
- In worksheet steps 6–9, remind students to press *Reset* to return all of the tiles to the stacks after each problem.

SUMMARIZE

Project the sketch. Expect to spend about 5 minutes.

9. Gather the class. Students should have their worksheets with them. Ask, ***How did you multiply polynomials today?*** Establish as a class that multiplying polynomials can be represented by the multiplication equation $length \times width = area$ (or, $base \times height = area$), where the dimensions correspond to the factors and the area corresponds to the product.
10. ***How did you decide which tiles to use to build a rectangle? How did you determine which tile goes in each place?*** Students should understand that every tile is a product of two dimension tiles, one on each side of the frame, and that they have to consider both dimensions when choosing their area tiles.
11. ***Are there shortcuts to multiplying? Do you always have to use tiles to multiply polynomials?*** Students should be able to make a multiplication table with dimensions that correspond to the factors and predict the quantity and type of tiles that belong in each region.
12. If time permits, discuss the Explore More. Let students show arrangements they made using the custom tools and how the arrangements adjust to the values of the sliders. You may consider giving all students the chance to work with the dynamic tile tools on another day.

EXTEND

1. Ask, ***What are some limitations of building rectangles with algebra tiles? Do the limitations still exist with multiplication tables?*** Limitations include the inability to represent negative values, nonlinear dimensions, or variables other than x and y, and the lack of enough pieces for large polynomials. Multiplication tables do not have these limitations.

2. Ask, ***If the product of two polynomials can be represented by the area of a rectangle, how could you represent the product of three polynomials?*** The product of three polynomials can be represented by the volume of a rectangular prism.

ANSWERS

Get Ready

A. $EF = 7$, $BC = 7$. The height of a rectangle is constant.

B. $2x^2 + xy + 3x + 2y + 3$

C. $2x + 2y + 4$; $4x + 4$; $4x + 2y + 6$

D. $2 \times 5 = 10$; $6 \times 7 = 42$; $a \times b = ab$

E.

3

4

F.

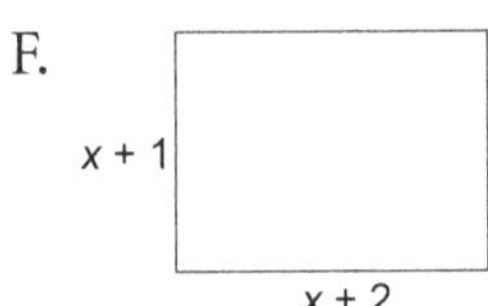

Answers for area will vary. Students should communicate that they understand that $x + 1$ and $x + 2$ should be multiplied.

Explore

4.

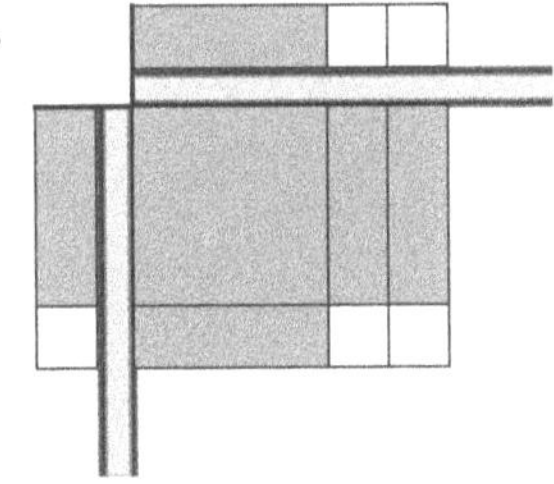

$(x + 1)(x + 2) = x^2 + 3x + 2$

6.

$(5)(x + 2) = 5x + 10$

7.

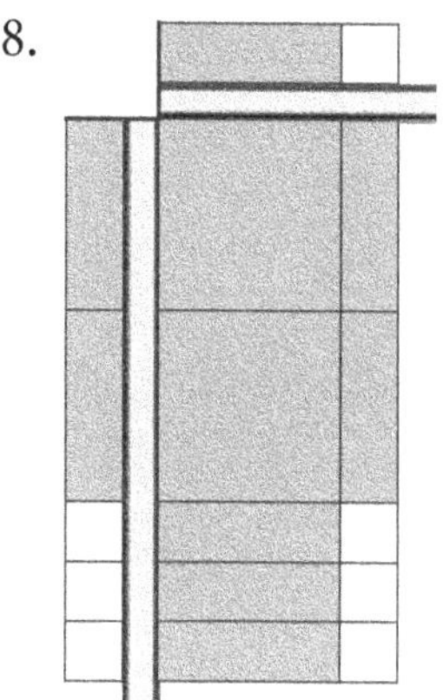

$(y + 2)(3x) = 3xy + 6x$

8.

$(2x + 3)(x + 1) = 2x^2 + 5x + 3$

9. 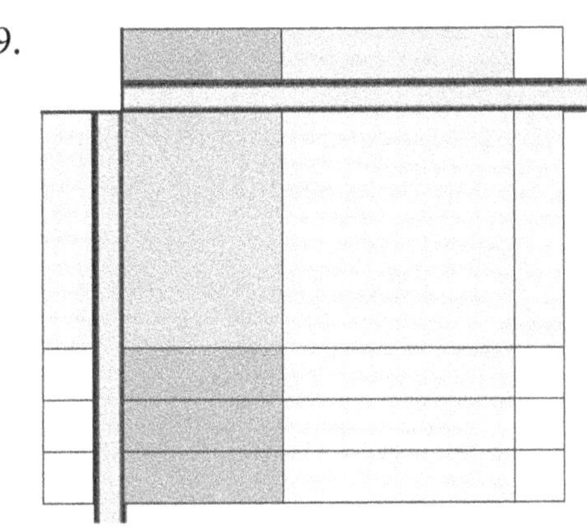

   $(y + 3)(x + y + 1) = xy + y^2 + 3x + 4y + 3$

10. Building the rectangle would require 40 tiles. (6 x^2 tiles, 22 x tiles, and 12 unit tiles)

11. The red points appear wherever like terms are separated in the dimension tile.

12.

	$3x$	2
$2x$	$6x^2$	$4x$
6	$18x$	12

   $(2x + 6)(3x + 2) = 6x^2 + 22x + 12$

13.

	x	y
$4x$	$4x^2$	$4xy$
7	$7x$	$7y$

   $(4x + 7)(x + y) = 4x^2 + 4xy + 7x + 7y$

14.

	x	y	1
y	xy	y^2	y
3	$3x$	$3y$	3

   $(y + 3)(x + y + 1) = xy + y^2 + 3x + 4y + 3$

   The number of rows and columns correspond to the number of terms in each of the factors.

Tiling in a Frame

For GSP5 Name:

GET READY

A. In the figure below, *ABCD* is a rectangle. Segment *EF* is perpendicular to segment *DC*. What are the measures of segments *EF* and *BC*? Why?

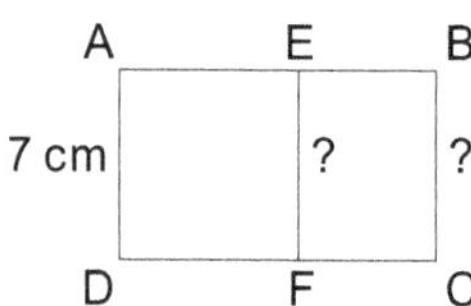

B. Write the following area collection as a sum. Combine like terms.

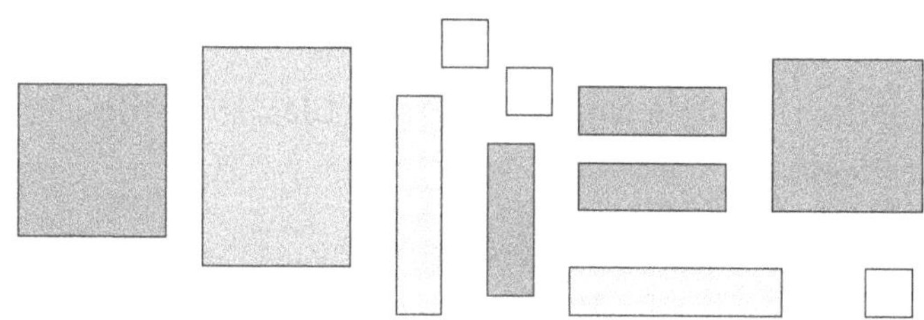

C. Find the perimeter of the following rectangles.

a. 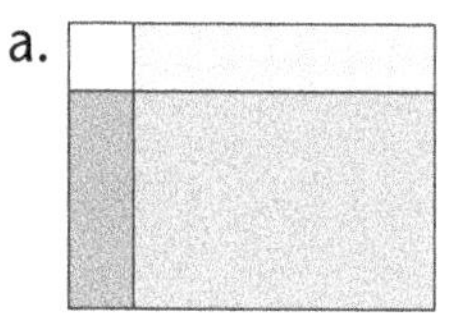

b. 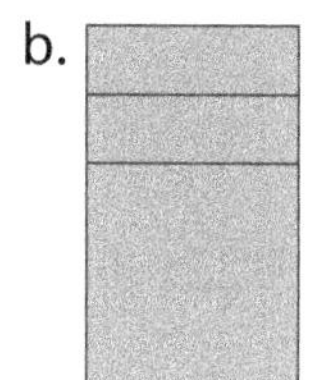

c. 

D. Find the area of the rectangles shown here. Write the area as a product of the dimensions.

2 in
5 in

____ ____ = _____
Dimensions Area

6 ft
7 ft

____ ____ = _____
Dimensions Area

a
b

____ ____ = _____
Dimensions Area

E. Draw a picture of a rectangle representing the product $(3)(4) = 12$.

F. Draw a picture of a rectangle representing the product $(x + 1)(x + 2)$. What do you think this area will be?

Tiling in a Frame

Name:

In this activity you'll explore the areas of rectangles when you know the dimensions, even if the length and width include variables.

EXPLORE

1. Open **Tiling in a Frame.gsp.** If you don't know the names of the tiles, go to page "Tiles." Otherwise, go to page "Frame."

2. Drag tiles from the stacks to the outside edges of the frame to represent the product $(x + 1)(x + 2)$.

3. Use tiles to build a rectangle inside the frame with dimensions that match those along the outside edges of the frame.

4. Make a sketch of the finished arrangement and write out the multiplication as an area equation.

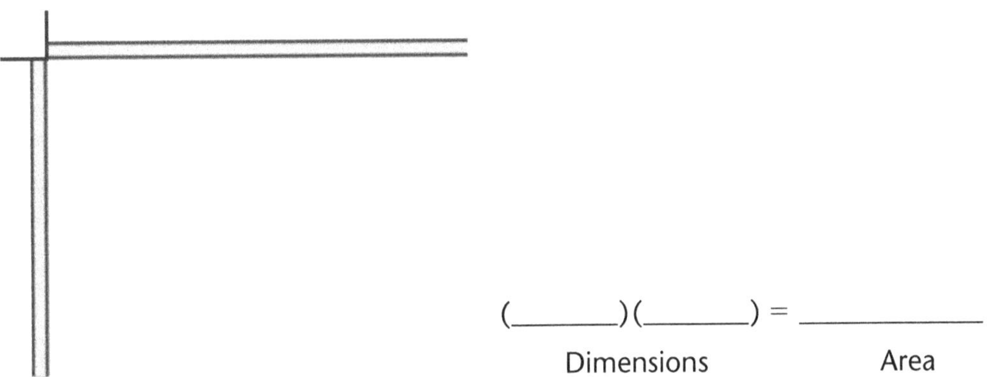

(_______)(_______) = ____________

Dimensions Area

5. Press *Reset* to return all tiles to the stacks.

6. Repeat steps 2–5 for the product $(5)(x + 2)$.

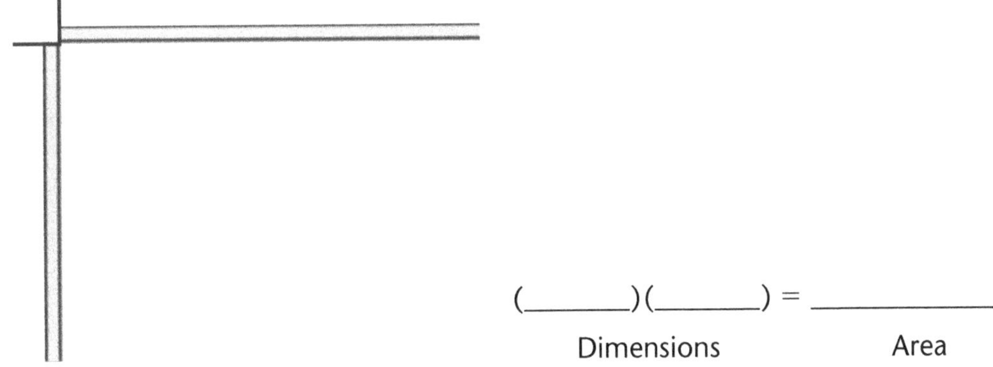

(_______)(_______) = ____________

Dimensions Area

7. Multiply $(y + 2)(3x)$. Sketch the finished arrangement.

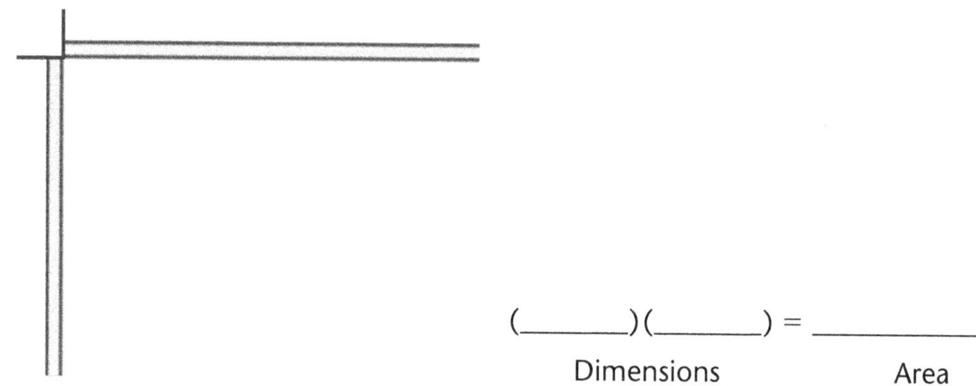

(_______)(_______) = ____________

Dimensions Area

8. Multiply $(2x + 3)(x + 1)$. Sketch the finished arrangement.

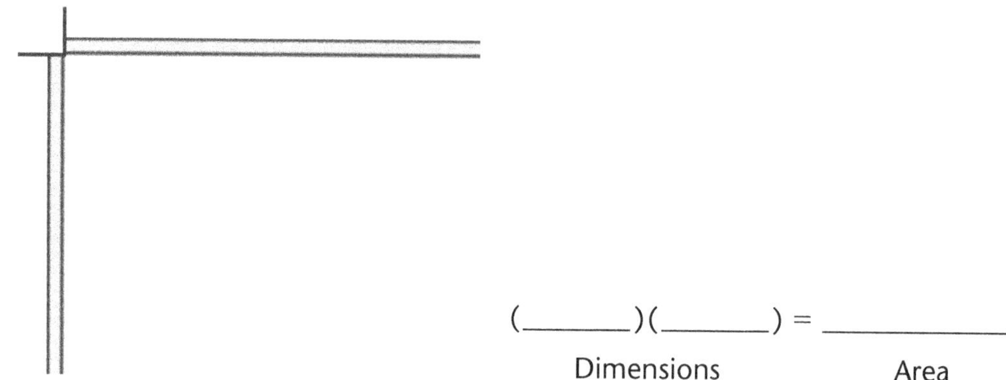

(_______)(_______) = ____________

Dimensions Area

9. Multiply $(y + 3)(x + y + 1)$. Sketch the finished arrangement.

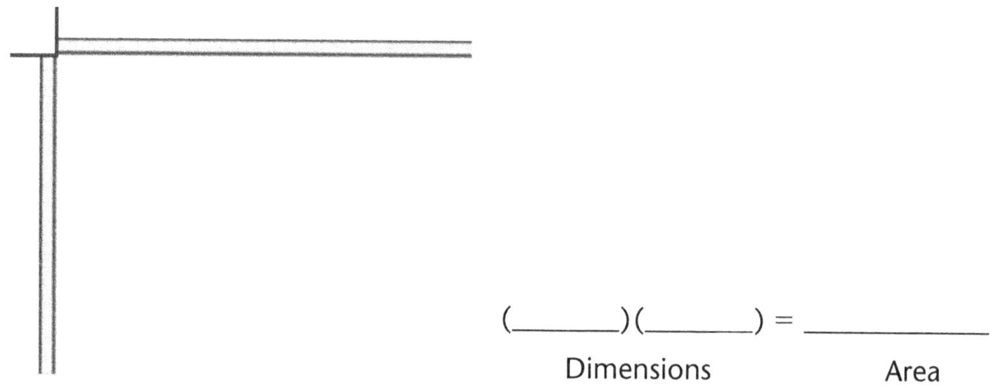

(_______)(_______) = ____________

Dimensions Area

10. Go to page "Regions." Suppose you had to multiply $(2x + 6)(3x + 2)$. How many tiles would you need to build a rectangle for this?

11. Press *Show Regions.* Look at the red points, where the dashed lines begin. How are these red points related to the dimension tiles?

12. Instead of building a rectangle, you'll divide the space into regions, separating like terms. Complete the multiplication table and write the multiplication equation.

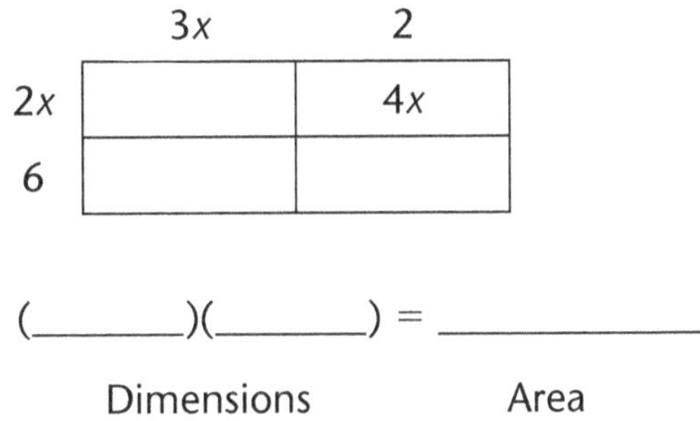

	$3x$	2
$2x$		$4x$
6		

(________)(________) = _____________

Dimensions Area

13. Multiply $(4x + 7)(x + y)$ as you did in step 12.

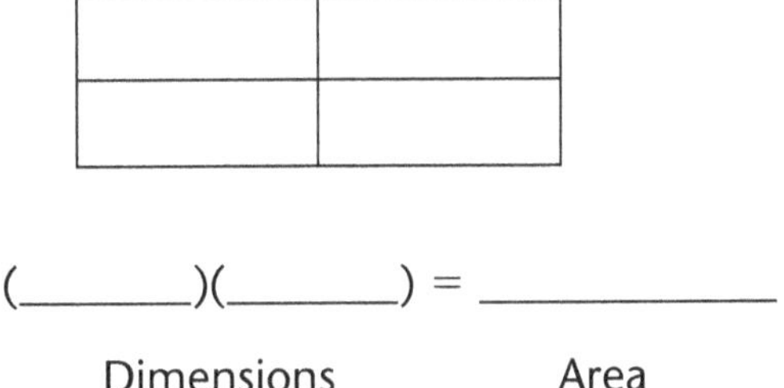

(________)(________) = _____________

Dimensions Area

14. Multiply $(y + 3)(x + y + 1)$ using a multiplication table. How can you determine the number of columns and rows you need in the table?

EXPLORE MORE

15. Go to page "Explore More." Use the dynamic tile tools in the Custom Tools menu to construct an arrangement. Choose a tool and click an existing point in the sketch. There are three points you can start with, one for each dimension along the frame and one for the inside of the frame. To connect new tiles to existing tiles, move the pointer until a point on the existing tile becomes highlighted and then click.

    Drag the x and y sliders to see how the arrangement adjusts to their new values. If you've constructed your arrangement correctly, it will hold together when you drag the sliders.

Tiling Rectangles: Factoring Polynomials

ACTIVITY NOTES

This activity can be used by students whether or not they have had hands-on experience with physical algebra tiles. However, students should have already completed the activity Tiling in a Frame, which introduces the Sketchpad algebra tiles.

If your students have experience with physical tiles, you might have them use custom tools to construct arrangements that will update dynamically as students drag the *x* and *y* sliders. If so, students should use page "Explore More," rather than page "Frame," to complete the worksheet. Students who do not have experience with physical tiles will need to work with the Sketchpad tiles before they can effectively use the custom tools.

INTRODUCE

Project the sketch for viewing by the class. Expect to spend about 10 minutes.

1. The first page of the student worksheet—Get Ready—can be done in class the day before students use computers, assigned as homework, or used as a warm-up. In any case, you should review the Get Ready questions before starting the activity. Ask for volunteers to explain their answers. Emphasize that the area of a rectangle can be expressed as the product of the dimensions and that the *x* and *y* tiles can be used to represent both area and dimension because they have a unit width.
2. Open **Tiling Rectangles.gsp** and go to page "Tiles." Remind students that tiles are named after their areas. Drag the *x* and *y* sliders to show that the area expressions match the dimensions for all values. Explain that this activity uses rigid tiles, but that students can explore dynamic tiles in the Explore More.
3. Go to page "Example 1" (or go to page "Frame" and make up your own examples). Enlarge the document window to fill the screen. Explain, ***Today you're going to use Sketchpad to make rectangular arrangements with tiles, but this time you won't be given the dimensions. Instead you'll be given the area and you'll need to find the dimensions. You'll do this by making a rectangle in the frame. For example, here is one way to arrange the expression $4x + xy + 6y + y^2 + 8$ into a rectangle. Notice that the tiles are arranged so that there are no gaps or overlaps.*** Press *Show Blueprint Lines.* Then press *Hide Area Tiles.* Explain that the blueprint lines also work in reverse. Students can check that the "seams" in the rectangle align with the "seams" along the outside of the frame.

4. Drag an inappropriate tile, such as an xy or a y^2, along the outside of the frame and say, ***It looks like this piece fits here. Is this correct?*** Students should understand that dimension tiles need to have a unit width, which in this model includes only the 1, x, and y tiles. The tiles along the outside of the frame should not extend beyond the black segments in the upper-left corner of the frame.

5. Press *Show Dimension Tiles.* ***What are the dimensions of the completed rectangle?*** $[(y + 4)$ and $(y + x + 2)]$ ***The dimensions represent the factors of the original polynomial. You can think of factoring as a multiplication problem in reverse.*** Use a text caption, or use the board, to write the multiplication problem.

$$area = length \times width$$

$$(\text{or, } area = base \times height)$$

$$4x + xy + 6y + y^2 + 8 = (y + 2)(y + x + 4)$$

6. Go to page "Example 2" and press *First Attempt.* ***This is my first attempt at making a rectangle for $x^2 + 7x + 10$. What do you think?*** Elicit the ideas that the tiles do not line up correctly and that the overall shape is not actually a rectangle. Press *Show Blueprint Lines* to show that some of the blueprint lines pass through the x^2 tile. Then press *Second Attempt.* You might clarify that some of the horizontal x tiles needed to be rotated to be vertical in order to make the rectangle, and that is fine. They are the same pieces and have the same area. ***Can you tell what the factors will be now?*** $[(x + 5)$ and $(x + 2)]$. Make sure students understand that the dimensions represent the factors of the polynomial.

DEVELOP

Expect students at computers to spend about 30 minutes.

7. Assign students to computers and tell them where to locate **Tiling Rectangles.gsp.** Distribute the worksheet. Tell students to work through step 15 and do the Explore More if they have time. Encourage students to ask their neighbors for help if they are having difficulty with Sketchpad.

8. Let pairs work at their own pace. As you circulate, here are some things to notice.

   - Make sure students follow the "no gaps or overlaps" rule when making their rectangles. The tiles are noncommensurate and a tile will not align unless it shares the same dimensions as the adjacent tile. You can use the idea that blueprint lines must run through the rectangle to the frame uninterrupted.

- Check that students use only the 1, x, and y tiles to represent dimensions. The x^2, xy, and y^2 tiles can only represent area.
- Some students might need to be assured that horizontal and vertical versions of the same tile are in fact interchangeable. They have the same area. You might connect this to the conservation of area or the commutative property of multiplication.
- Although students can find many ways to make a rectangle, some students might find it easier to always group similar tiles together, with the larger tiles in the upper-left corner and the unit tiles in the lower-right corner. This facilitates trinomial factoring by showing the factors of the constant term as rectangular dimensions.
- Tiles can be dragged back to the stacks if a student decides that a piece is not needed. Students can also use **Edit | Undo** to return the last piece, or they can press *Reset* to return all tiles to the stacks.
- In worksheet steps 8 and 9, there are two ways to factor each of these polynomials.

SUMMARIZE

Project the sketch. Expect to spend about 5 minutes.

9. Gather the class. Students should have their worksheets with them. Ask, ***Today you factored polynomials. What does it mean to factor a polynomial?*** Establish as a class that factoring is the opposite process of multiplying. You start with the product and need to find two factors that, when multiplied, give the product.
10. ***How can the area of a rectangle be written as a sum and a product?*** Students should be able to identify that the area of a rectangle is the sum of the parts, which is equivalent to the product of the dimensions. Factoring can be represented by the multiplication equation *area* = *length* × *width* (or, *area* = *base* × *height*), where the area corresponds to the product and the dimensions correspond to the factors.
11. ***Does the order of the factors change the product?*** Students should be able to justify the commutative property of multiplication as evidenced by the conservation of the area tiles.

12. ***How can we use the factors of the last term of the polynomial to make the rectangular arrangement?*** Students should be able to identify a means of using the constant term factors to determine the grouping of the x tiles.

13. If time permits, discuss the Explore More. Let students show arrangements they made using the custom tools and how they adjust to the values of the sliders. You may consider giving all students the chance to work with the dynamic tile tools on another day.

14. You may wish to have students respond individually in writing to this prompt: ***Explain how to find the factors of a rectangle whose area tiles are $x^2 + 8x + 7$. Explain why this polynomial has only one pair of factors. Give an example of a rectangle that has more than one factor pair.***

EXTEND

Ask, ***Will this factoring process work for polynomials that have more than one x^2? For example, what are the factors of $2x^2 + 3x + 1$?*** [$(2x + 1)$ and $(x + 1)$]. Create more challenging factoring problems, such as $3x^2 + 20x + 12 = (3x + 2)(x + 6)$. Remind students that they can work this process in reverse. For instance, you might first ask, ***What is the product of $(y + 1)(2x + 3)$?***

ANSWERS

Get Ready

A. 1, 6 and 2, 3

B.

6

1 6

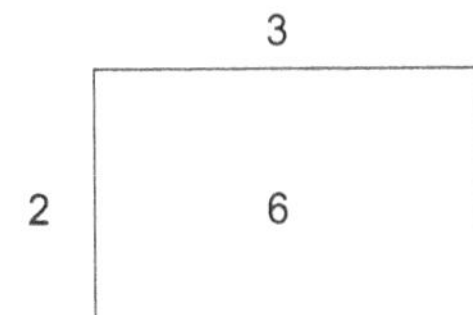

The 6 should be placed inside because it is counting the interior squares.

C. Area = 12.7

D.

Type	Quantity
1	8
x	4
y	6
xy	1
x^2	0
y^2	1

E. Area is $4x + xy + 6y + y^2 + 8$; dimensions are $(y + 4)$ and $(y + x + 2)$.

F. $x^2 + 7x + 10$

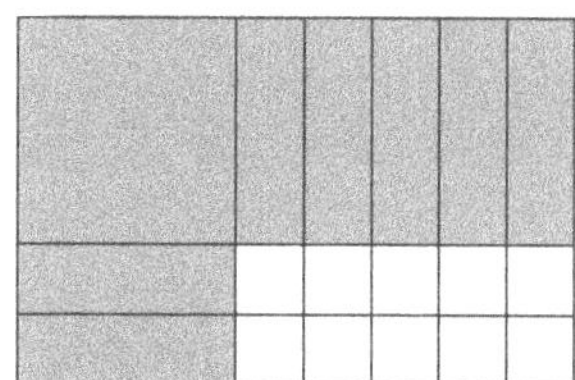

Dimensions are $(x + 2)$ and $(x + 5)$.

Explore

4.

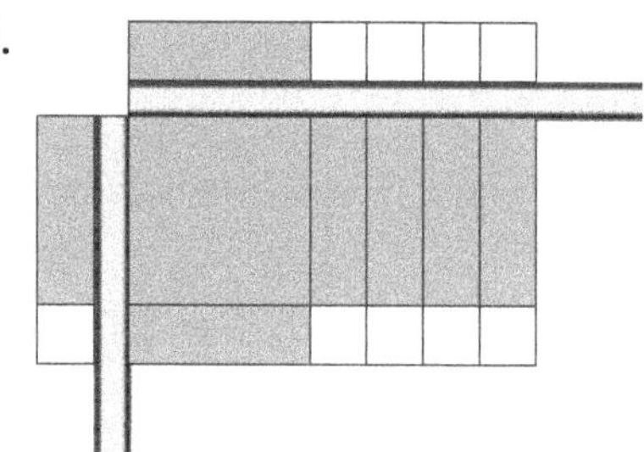

$x^2 + 5x + 4 = (x + 4)(x + 1)$

5.

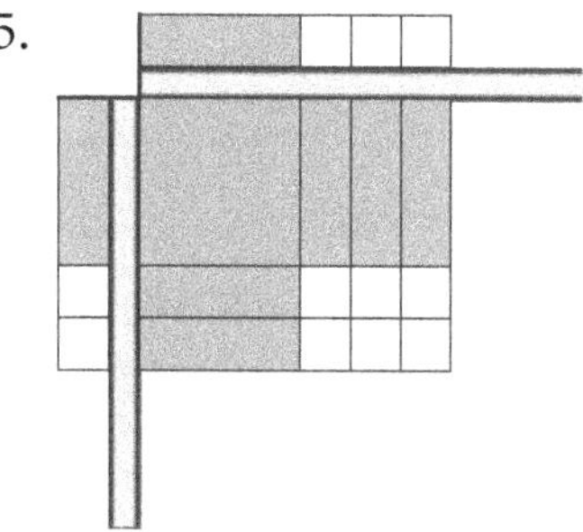

$x^2 + 5x + 6 = (x + 3)(x + 2)$

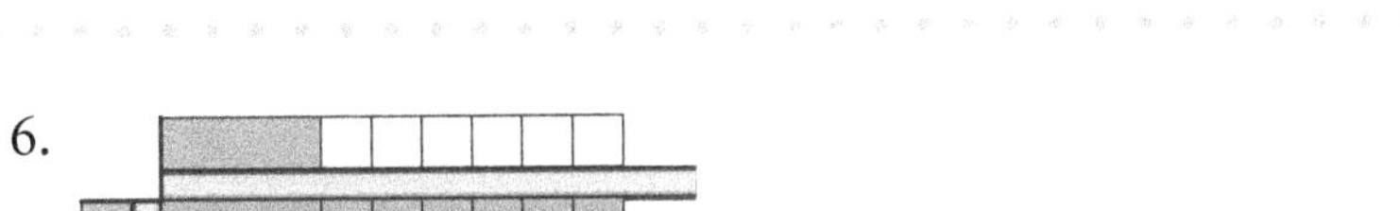

6.

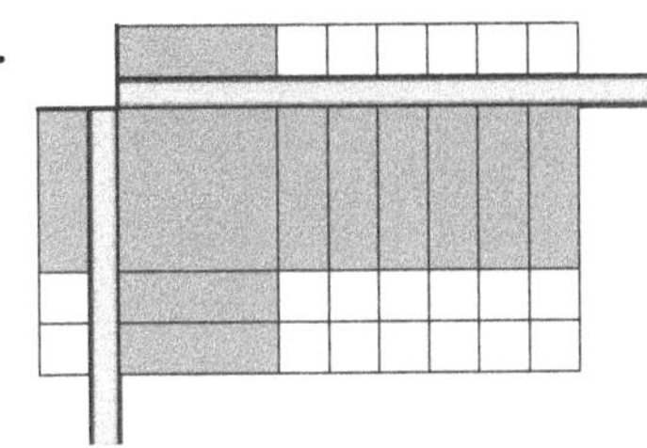

$x^2 + 8x + 12 = (x + 6)(x + 2)$

7.

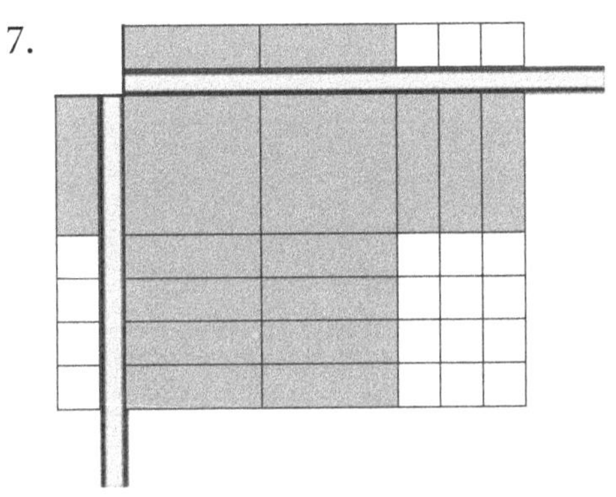

$2x^2 + 11x + 12 = (2x + 3)(x + 4)$

8.

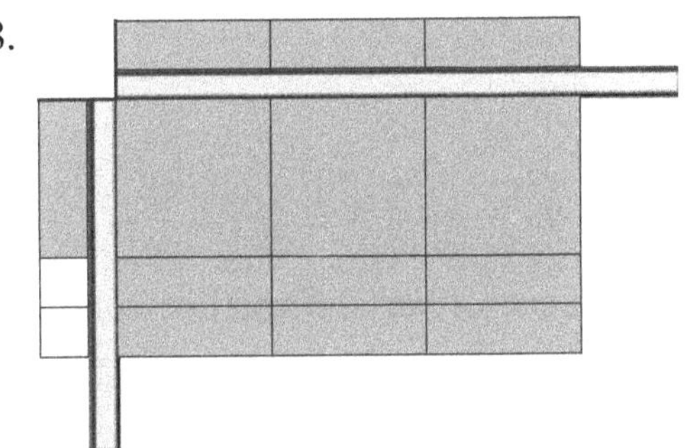

$3x^2 + 6x = (3x)(x + 2)$ (or, $3x^2 + 6x = (x)(3x + 6)$)

9.

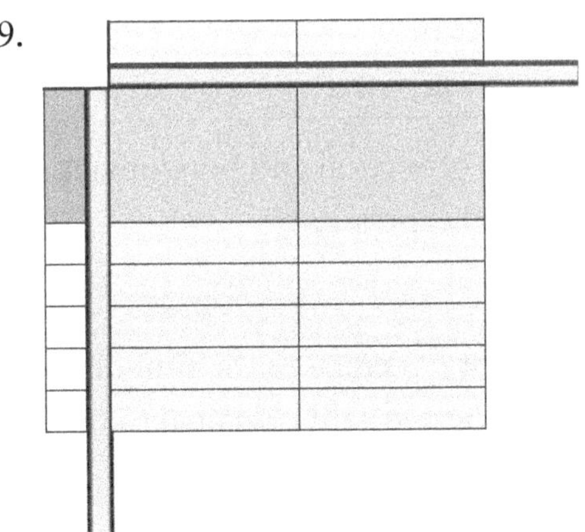

$2xy + 10y = (2y)(x + 5)$ (or, $2xy + 10y = (y)(2x + 10)$)

10. 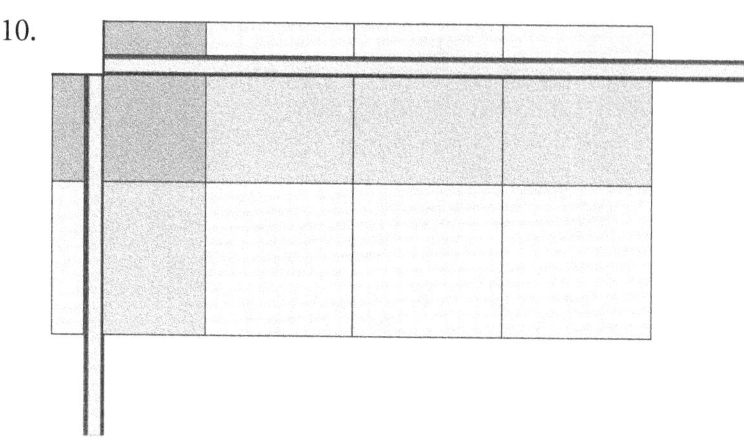

$x^2 + 4xy + 3y^2 = (x + y)(x + 3y)$

11. 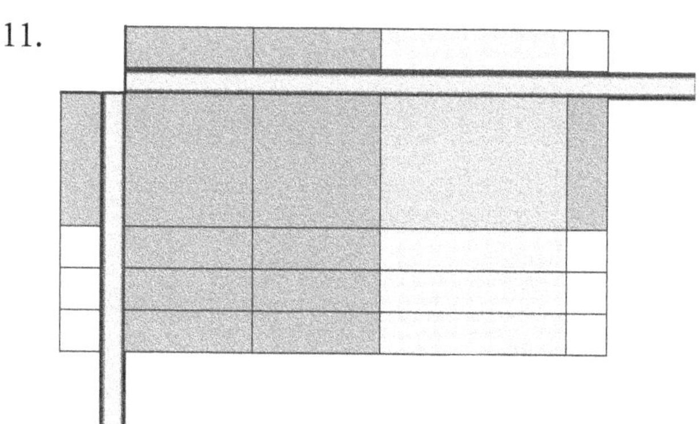

$2x^2 + xy + 7x + 3y + 3 = (2x + y + 1)(x + 3)$

12. *Note:* Numbers for vertical and horizontal *x* tiles can be reversed.

Problem	Polynomial	Number of Vertical *x* Tiles	Number of Horizontal *x* Tiles
Step 4	$x^2 + 5x + 4$	4	1
Step 6	$x^2 + 5x + 6$	3	2
Step 7	$x^2 + 8x + 12$	6	2

13. The numbers of vertical and horizontal tiles are equal to the constants of the factored expression.

14. $x^2 + 7x + 12 = (x + 3)(x + 4)$

15. The coefficient b is equal to the sum of the constants of the factored expression. The constant c is equal to the product of the constants in the factored expression.

Tiling Rectangles

For GSP5 Name:

GET READY

A. Factoring means finding the multipliers that make the product. The factor pairs of 4 are 1, 4 and 2, 2. What are the factor pairs of 6?

B. For each of the factor pairs you listed above, draw and label a rectangle that has the same dimensions as your factor pair. Where should the value 6 be placed for all of your rectangles? Why?

C. Notice that any number will have a factor pair that includes the number 1. Does the value 1 show up in the area? Find the area of the rectangle.

12.7

1

D. List the number of each type of tile in the arrangement.

Type	Quantity
1	
x	
y	
xy	
x^2	
y^2	

E. What is the area of the arrangement above?
What are the dimensions?

F. What polynomial does the arrangement shown here represent? Can you arrange the pieces into a rectangle? If so, what are its dimensions?

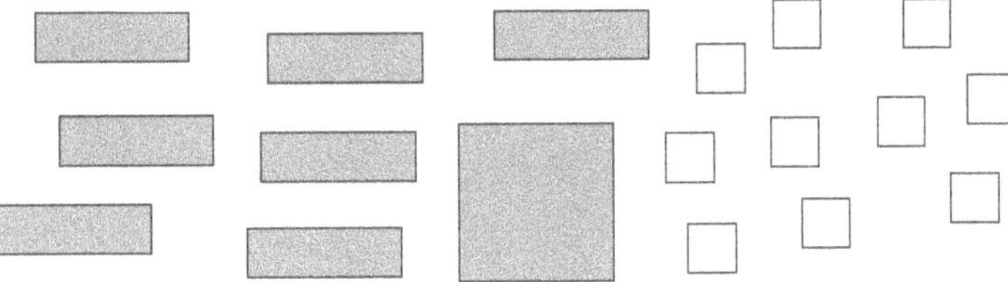

Tiling Rectangles

Name:

In this activity you'll use algebra tiles to make rectangles and find their dimensions as a model for factoring polynomials.

EXPLORE

1. Open **Tiling Rectangles.gsp** and go to page "Frame."

2. Drag tiles from the stacks to the inside of the frame to represent the polynomial $x^2 + 5x + 4$. Arrange the tiles into a rectangle. You can exchange horizontal and vertical x tiles as needed.

3. Drag tiles to the outside of the frame to represent the dimensions of the rectangle.

4. Sketch the finished arrangement and express it as an area equation.

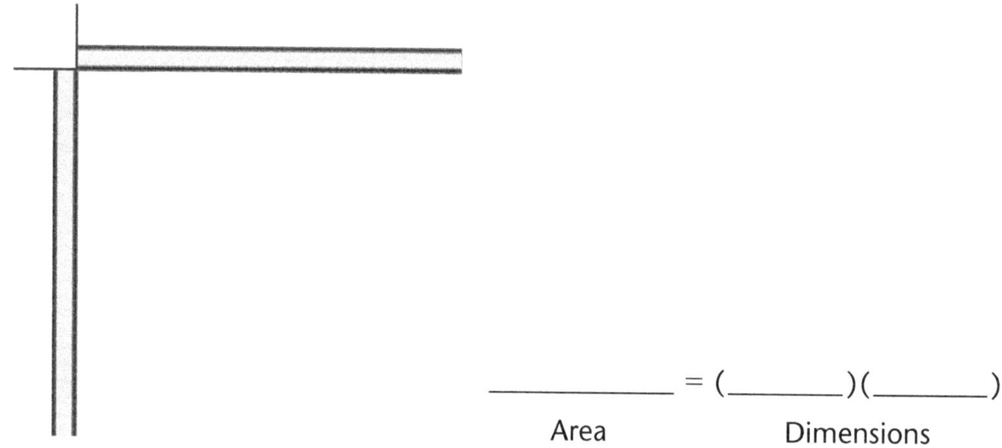

________ = (______)(______)

Area Dimensions

5. Press *Reset.* Repeat steps 2–4 for the polynomial $x^2 + 5x + 6$.

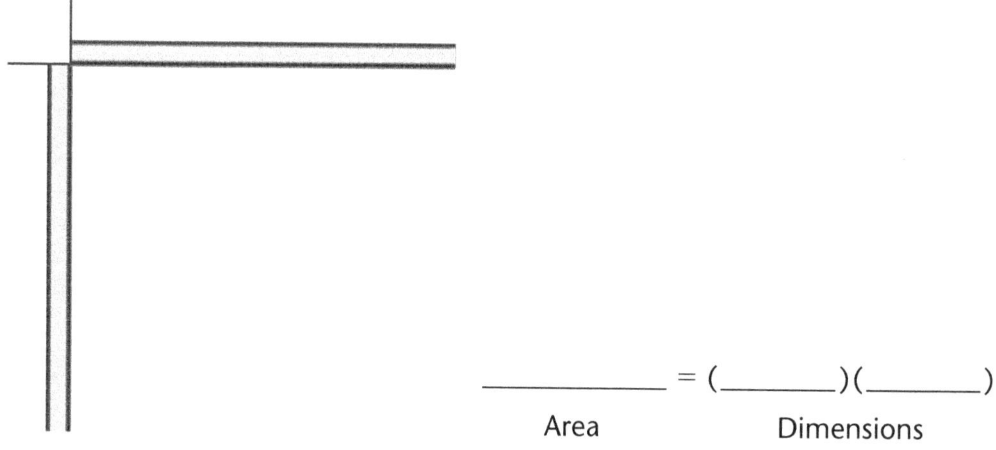

________ = (______)(______)

Area Dimensions

6. Factor $x^2 + 8x + 12$. Sketch the finished arrangement.

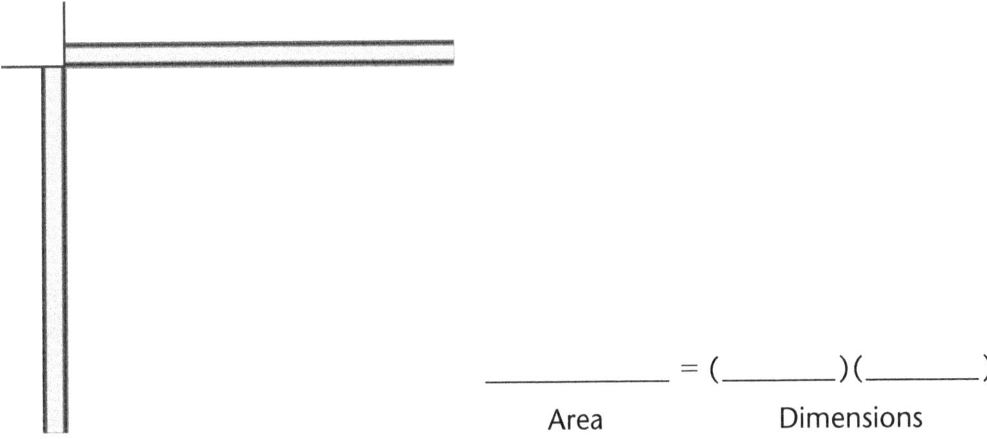

__________ = (________)(________)
Area Dimensions

7. Factor $2x^2 + 11x + 12$. Sketch the finished arrangement.

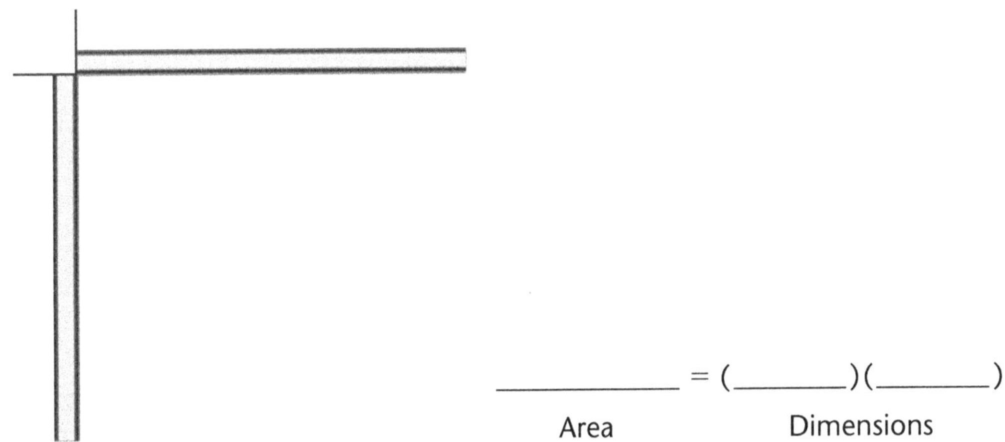

__________ = (________)(________)
Area Dimensions

8. Factor $3x^2 + 6x$. Sketch the finished arrangement.

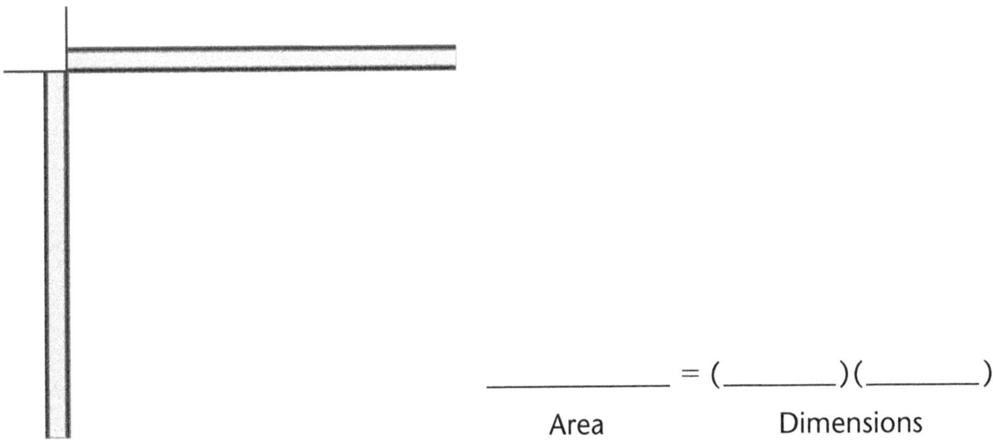

__________ = (________)(________)
Area Dimensions

9. Factor $2xy + 10y$. Sketch the finished arrangement.

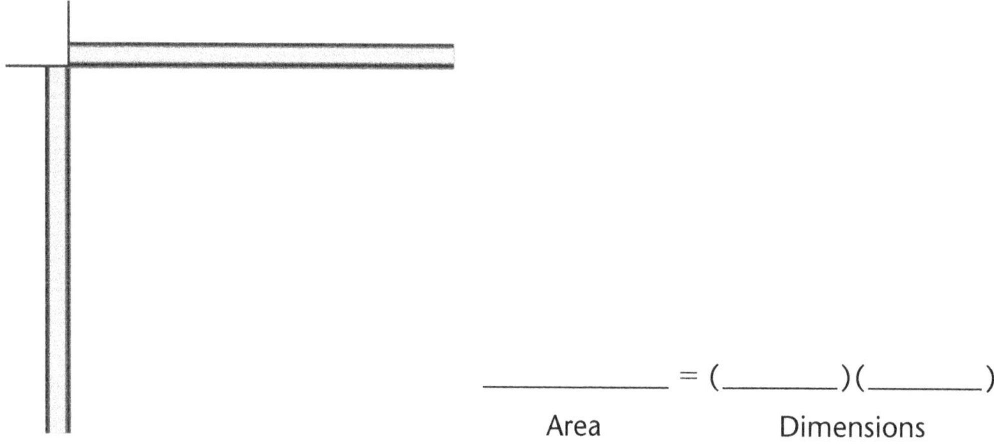

10. Factor $x^2 + 4xy + 3y^2$. Sketch the finished arrangement.

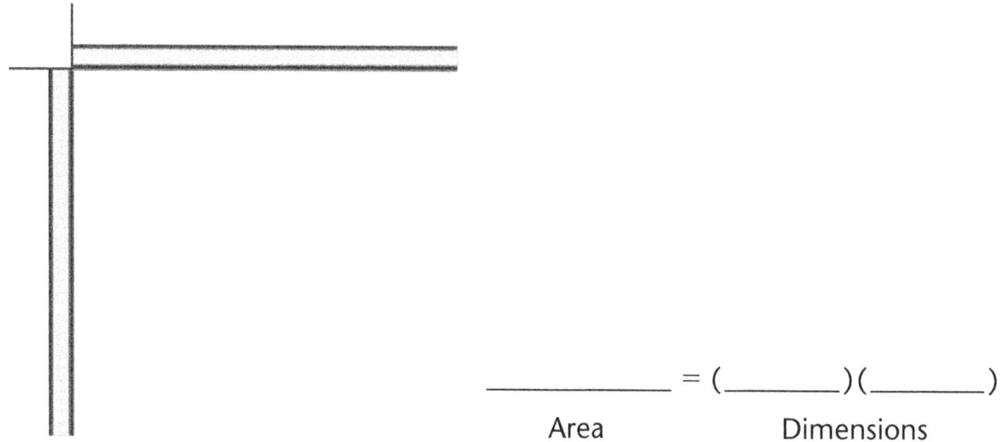

11. Factor $2x^2 + xy + 7x + 3y + 3$. Sketch the finished arrangement.

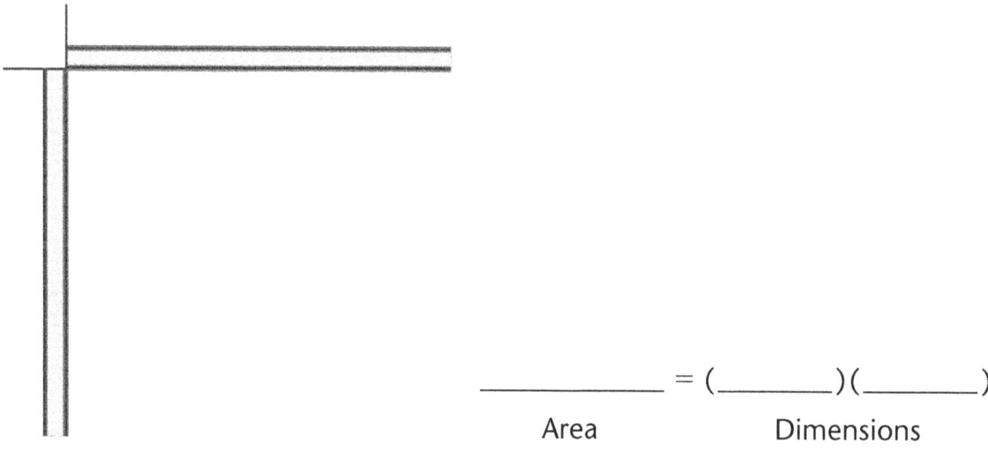

12. Look back at steps 4, 5, and 6. In each of these problems, there is only one x^2 tile, and some number of x tiles and 1 tiles.

    For each problem, list how many x tiles are vertical and how many are horizontal in the rectangle inside the frame.

Problem	Polynomial	Number of Vertical *x* Tiles	Number of Horizontal *x* Tiles
Step 4	$x^2 + 5x + 4$		
Step 5	$x^2 + 5x + 6$		
Step 6	$x^2 + 8x + 12$		

13. How are the numbers of vertical and horizontal x tiles related to the dimensions of the rectangle?

14. Use your observation in step 13 to find the factors of $x^2 + 7x + 12$ without using the Sketchpad model.

15. For polynomials of the form $x^2 + bx + c$, how are the factors related to the values of b and c?

EXPLORE MORE

16. Go to page "Explore More." Use the dynamic tile tools in the Custom Tools menu to construct an arrangement. Choose a tool and click on an existing point in the sketch. There are three points you can start with, one for the inside of the frame and one for each dimension along the frame. To connect new tiles to existing tiles, move the pointer until a point on the existing tile becomes highlighted and then click.

    Drag the x and y sliders to see how the arrangement adjusts to their new values. If you've constructed your arrangement correctly, it will hold together when you drag the sliders.

Dynamic Tiles: Evaluating Polynomial Expressions

ACTIVITY NOTES

This activity can be used by students whether or not they have had hands-on experience with algebra tiles. Most students, especially tactile learners, benefit from manipulating physical tiles. This activity, however, includes sliders that allow students to change the lengths of tiles with variable dimensions, which can't be modeled with physical tiles.

This activity can be completed in one extended 75-minute session or in two shorter sessions. If you wish to complete the activity in one 60-minute session, you might have the students do the Explore More on another day. The Explore More, which does not require the use of custom tools, can be done separately from the rest of the activity.

INTRODUCE

Project the sketch for viewing by the class. Expect to spend about 10 minutes.

1. Open **Dynamic Tiles.gsp.** Enlarge the document window to fill the screen.
2. If your students don't have prior hands-on experience with algebra tiles, go to page "Tiles." Make sure that the dimensions are showing. Explain that tiles are named after their areas. Make sure students can identify all their names. Then press *Show Areas.* Drag the x and y sliders to show that the area expressions match the dimensions for all values. If your students are already familiar with algebra tiles, you can skip this step.
3. Go to page "(A) xy." Say, ***Today you're going to use Sketchpad to evaluate polynomial expressions. You'll use different values for the variables x and y to see how the expressions change. Before you start, I'll show you how to use the model.***
4. Make sure the grid is showing. Drag the sliders so that $x = 8$ and $y = 6$. Ask, ***What is the area of the light blue rectangle?*** [48] Drag the x slider to 3. ***What is the area now?*** [18] Drag the y slider to 4 and ask again. [12] Then press *Hide Grid.* ***What are the names of these three tiles?*** [Purple is x, green is y, and blue is xy.] Press *Show Names.*
5. Go to page "(B) x(2y)." Press *Show Variable Tiles* and ask for the names again. Ask, ***How can I write an equation using these names to say that the length multiplied by the width is equal to the area?*** Use a text caption, or use the board, to write the multiplication problem.

$$(x)(2y) = 2xy$$

6. Drag the sliders so that $x = 2$ and $y = 3$. Ask, ***What is the area of the rectangular region inside the frame when x is 2 and y is 3?*** [12] Let students respond, and then press *Show Unit Tiles.*

7. Drag the *x* slider so that $x = 5$. Explain how to use the custom tools. ***In some parts of this activity, you'll use custom tools to add more tiles to the sketch. If I want to use unit tiles to fill in the rectangle when x is 5 and y is 3, I click on the* Custom *tool icon and choose the* 1 *tool. Then I click on the sketch to create more unit tiles. In order for the new tiles to be connected to the tiles that are already there, I click on the points at the corners of the existing tiles. Notice how the point becomes highlighted when I'm at the right spot.*** Let students know that the custom tool remains active until they select the **Arrow** tool (or any other tool).

8. Model how to add the unit tiles, going from left to right as you add each row, until there are 30 tiles. Remind students that they can always use **Edit | Undo** if they make a mistake. You might mention the difference between the horizontal and vertical versions of the *x*, *y*, and *xy* tiles, and that x^2 means x^2.

9. If you want students to save their work, demonstrate choosing **File | Save As,** and let them know how to name and where to save their files.

DEVELOP

Expect students at computers to spend about 60 minutes.

10. Assign students to computers and tell them where to locate **Dynamic Tiles.gsp.** Distribute the worksheet. Tell students to work through step 30 and do the Explore More if they have time. Encourage students to ask their neighbors for help if they are having difficulty with Sketchpad.

11. Let pairs work at their own pace. As you circulate, here are some things to notice.

    - In worksheet step 1, when $x = 11$ and $y = 12$, students should be encouraged to use multiplication instead of counting units.
    - In worksheet step 11, help students as needed to correctly use the **1** custom tool. Move the tile into the frame, but do not click until the point on the neighboring tile is highlighted.
    - When transitioning from adding tiles with a custom tool to dragging the sliders, remind them to choose the **Arrow** tool (or any other tool) to deselect the custom tool.

- In worksheet step 13, the area formed by the unit tiles will not remain 42, but the students should witness that there are 18 unit tiles that are no longer part of the evaluation for $x = 4$ and $y = 3$.
- In worksheet step 18, students may need clarification on how to fill in the area in the frame using the custom tool.
- In worksheet step 21, students are asked to use unit tiles to fill in the area in the frame. These tiles will overlap with the y tiles and should be used to justify their algebraic evaluation answers.
- In worksheet step 24, students are told to use the **x (vertical)** tool to add a tile to the dimension on the left side along the outside of the frame. They must then figure out to use the **xy (y horizontal)** tool to fill in the rectangle inside the frame.
- In worksheet step 30, students are discouraged from using unit tiles to count the area. You might ask about the limitation of the counting, and make sure they are correctly evaluating expressions algebraically.

12. If students will save their work, remind them where to save it now.

SUMMARIZE

Expect to spend about 5 minutes.

13. Gather the class. Students should have their worksheets with them. If students have completed the Explore More, have them share their solutions from worksheet step 34. Otherwise, review steps 26–30. Ask for volunteers to demonstrate the algebraic steps that verify their answers. Discuss the two methods of evaluating these rectangular areas: the product of the dimensions, or the sum of the areas. Both require adherence to the order of operations.

EXTEND

What questions might you ask about evaluating expressions? Encourage curiosity. Here are some sample student queries.

What does it mean to be a variable?

Can x and y have values that are negative?

How many values are there for x and y?

Are there any tiles that never change?

What is the difference between an expression and an equation?

ANSWERS

1.

x Slider Value	*y* Slider Value	Area of Rectangle in Frame
8	6	48
2	6	12
3	2	6
11	12	132
x	*y*	*xy*

2. 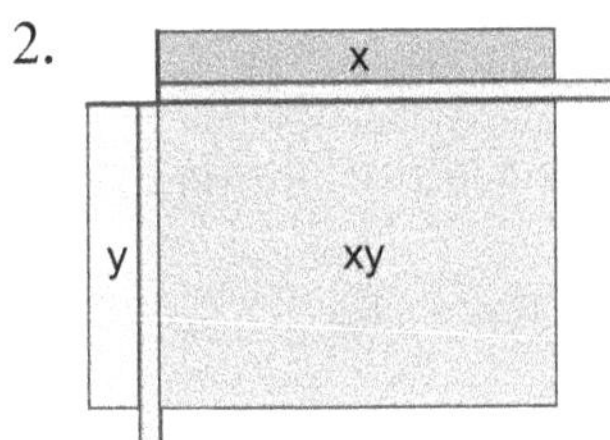

3. 12

4. 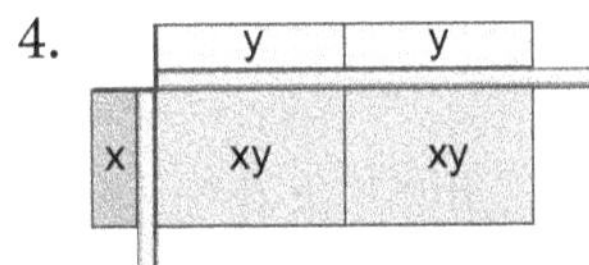

5. $2xy$

6. x and $2y$

7. $(x)(2y) = 2xy$

9. The tiles became longer vertically. The area is now more than 12.

11. 42

12. $7(2 \cdot 3) = 2 \cdot 7 \cdot 3$

    $7(6) = 14 \cdot 3$

    $42 = 42$

13. The area should now be less than 42 because the dimension tile on the left became shorter.

14. $4(2 \cdot 3) = 2 \cdot 4 \cdot 3$

    $4(6) = 8 \cdot 3$

    $24 = 24$

15. Answers will vary.

16. $2(2 \cdot 7.5) = 2 \cdot 2 \cdot 7.5$

$$2(15) = 4 \cdot 7.5$$

$$30 = 30$$

The light blue rectangle made up of two *xy* tiles has more area.

17.

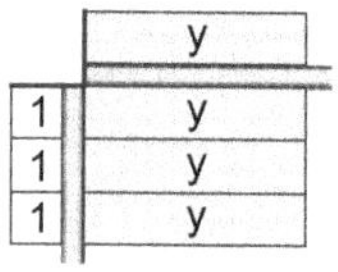

18. $(3)(y) = 3y$

19. Nothing happens to the picture because there are no x-values.

20. The rectangular region gets bigger or smaller because all the y-values change together.

21. $3(3) = 9$

22. $3(6) = 18$

23. $3(9) = 27$

24. An *xy* tile (*y* horizontal) will fill in the area in the frame.

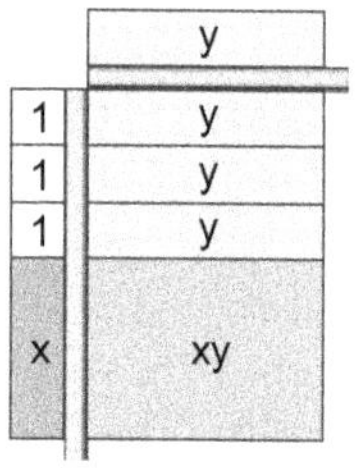

25.

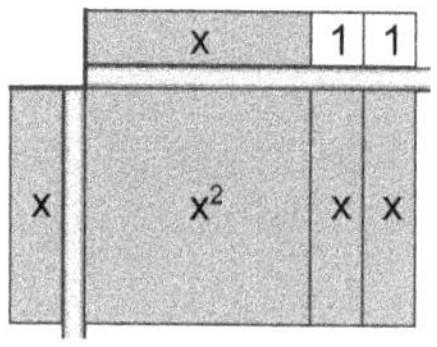

26. $(x)(x + 2) = x^2 + 2x$

27. $2(2 + 2) = 2^2 + 2 \cdot 2$

$$2(4) = 4 + 4$$

$$8 = 8$$

28. $4(4 + 2) = 4^2 + 2 \cdot 4$

$4(6) = 16 + 8$

$24 = 24$

29. $5(5 + 2) = 5^2 + 2 \cdot 5$

$5(7) = 25 + 10$

$35 = 35$

30. $11(11 + 2) = 11^2 + 2 \cdot 11$

$11(13) = 121 + 22$

$143 = 143$

31.

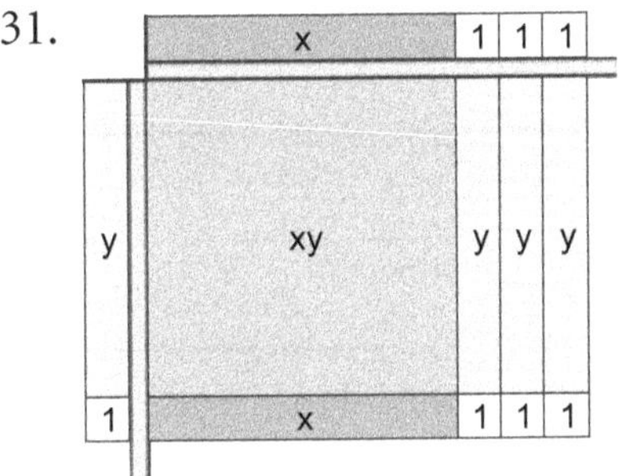

32. $(x + 3)(y + 1) = xy + x + 3y + 3$

33. $(9 + 3)(11 + 1) = 9 \cdot 11 + 9 + 3 \cdot 11 + 3$

$(12)(12) = 99 + 9 + 33 + 3$

$144 = 144$

34. Answers will vary. Sample solutions are shown in the tables.

12

x	*y*
3	1
1	2
0	3
9	0
2	1.4

16

x	*y*
5	1
1	3
13	0
2	2.2
7	0.6

30

x	*y*
2	5
7	2
0	9
3	4
1	7.5

36

x	*y*
1	8
0	11
9	2
3	5
6	3

35. Answers will vary. Sample solutions are shown in the tables above.

Dynamic Tiles

In this activity you will use algebra tiles to evaluate polynomial expressions. See whether you can identify different methods for finding the area in the frame.

EXPLORE

1. Open **Dynamic Tiles.gsp** and go to page "(A) xy." Drag the *x* and *y* sliders to the values shown in the table and find the area of the light blue rectangle inside the frame.

x **Slider Value**	*y* **Slider Value**	**Area of Rectangle in Frame**
8	6	
2	6	
3	2	
11	12	
x	*y*	

2. Press *Hide Grid.* Draw a diagram of the tiles inside the frame and along the outside of the frame. Label each tile with its name. Then press *Show Labels* to check.

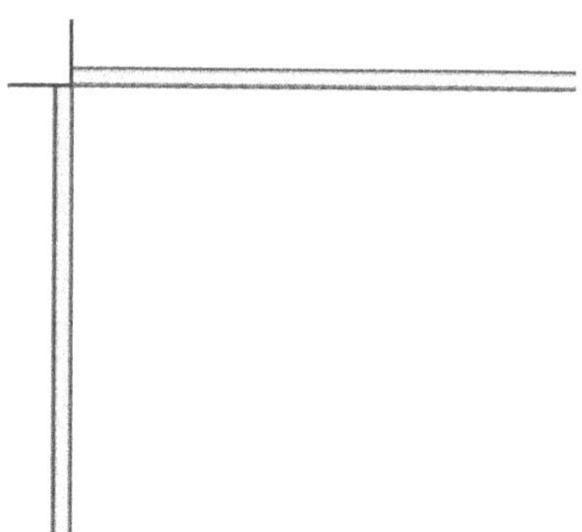

3. Go to page "(B) x(2y)." If needed, drag the *x* slider to 2 and the *y* slider to 3. What is the area of the yellow rectangle in the frame?

4. Press *Show Variable Tiles.* Draw a diagram of all the tiles and label them.

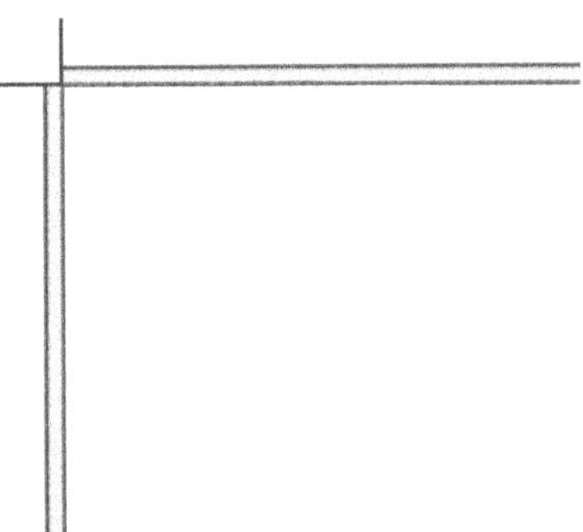

5. What is the area of the rectangle inside the frame?

6. What are the dimensions (length and width) represented by tiles along the outside of the frame?

7. Write the area equation shown by your diagram.

   (________) (________) = ______________

8. To evaluate the expressions on both sides of the area equation, you replace the variables with their values and simplify. The diagram in step 4 shows that $x(2y) = 2xy$. In step 3, when $x = 2$ and $y = 3$,

$$2(2 \cdot 3) = 2 \cdot 2 \cdot 3$$
$$2 \cdot (6) = 4 \cdot 3$$
$$12 = 12$$

9. Change the x-value to 7. How did the tiles along the outside of the frame change? Is the area inside the frame more or less than 12?

10. Press *Show Unit Tiles.* Now you'll add unit tiles until you fill in the area of the rectangle when $x = 7$ and $y = 3$.

11. Choose the **1** tool from the Custom Tools menu. Click the points along the bottom of the yellow squares starting from the left. Add enough unit tiles to fill in the area in the frame. How much area is in the frame?

12. Follow the process shown in step 8 to evaluate both sides of the equation when $x = 7$ and $y = 3$.

$$x(2y) = 2xy$$

$$\_\_\_(2 \cdot \_\_\_) = 2 \cdot \_\_\_ \cdot \_\_\_$$

$$\_\_\_(\_\_\_) = \_\_\_ \cdot \_\_\_$$

$$\_\_\_ = \_\_\_$$

13. Change the x-value to 4. Should the area determined by the new dimensions be more or less than 42? Why?

14. Evaluate both sides of the equation when $x = 4$ and $y = 3$.

15. Press *Show Variable Tiles.* Change the x-value to 2 and the y-value to 7.5. You should now see two rectangles, a light blue one made up of two xy tiles, and a yellow one made up of the unit tiles you added in step 11. Which one do you think has more area?

16. Find out by evaluating both sides of the equation $x(2y) = 2xy$ when $x = 2$ and $y = 7.5$.

17. Go to page "(C) 3y." Draw a diagram of the tiles along the outside of the frame and label them.

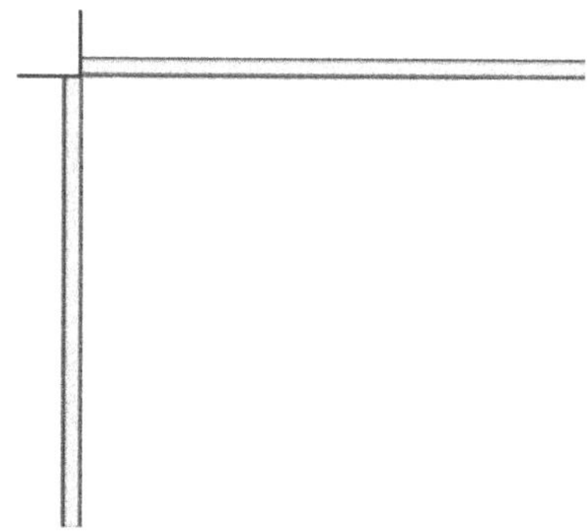

18. Choose the **y (horizontal)** tool. Fill the area inside the frame with *y* tiles. Add them to your diagram above and label them. Write the area equation shown by your diagram.

    (________) (________) = ______________

19. Drag the *x* slider. What happens to the sketch? Why?

20. Drag the *y* slider. What happens to your sketch? Why?

21. Change the value of *y* to 3. Use the **1** tool to fill in the area inside the frame with unit tiles. Evaluate the expression $3y$ when $y = 3$.

22. Change the value of *y* to 6. Use the **1** tool to fill in the area inside the frame with unit tiles. Evaluate the expression $3y$ when $y = 6$.

23. Change the value of *y* to 9. Use the **1** tool to fill in the area inside the frame with unit tiles. Evaluate the expression $3y$ when $y = 9$.

24. Choose the **x (vertical)** tool and add an *x* tile to the dimension on the left side along the outside of the frame. Without using unit tiles, which tile will complete the rectangle in the frame? Use the appropriate custom tool and complete the rectangle in the frame.

25. Go to page "(D) x(x+2)." Draw a diagram of the tiles along the outside of the frame and label them.

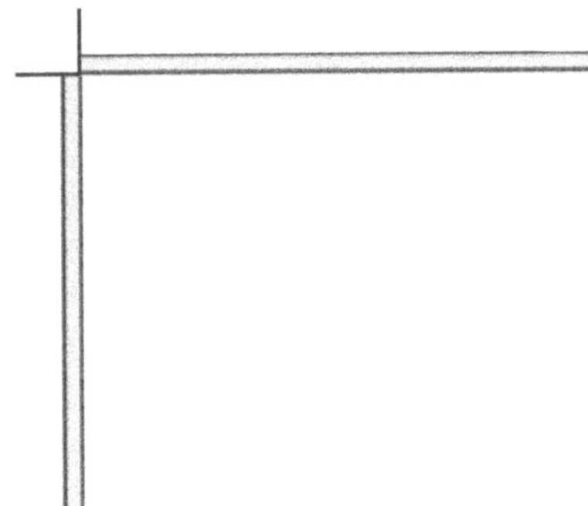

26. Fill in the area inside the frame without using any unit tiles. Choose the appropriate tools from the Custom Tools menu and add them to your sketch. Add them to your diagram above and label them. Write the area equation shown by your diagram.

    (________) (________) = ________________

27. Set x equal to 2. Fill in the area frame with unit tiles to count the area. Evaluate the expressions on both sides of the area equation for $x = 2$.

$$x(x + 2) = x^2 + 2x$$
$$\_\_\_(\_\_\_ + 2) = \_\_\_^2 + 2 \cdot \_\_\_$$
$$\_\_\_(\_\_\_) = \_\_\_ + \_\_\_$$
$$\_\_\_ = \_\_\_$$

28. Repeat step 27 for $x = 4$.

29. Repeat step 27 for $x = 5$.

30. Repeat step 27 for $x = 11$ (but don't add tiles anymore!).

EXPLORE MORE

31. Go to page "(E) (x+3)(y+1)." Draw a diagram of all the tiles and label them.

32. Write the area equation shown by your diagram.

    (________)(________) = ______________

33. Drag the sliders so $x = 9$ and $y = 11$. Evaluate the expressions on both sides of the area equation in step 32 for $x = 9$ and $y = 11$. Then press *Show Grid* to compare your solution with the diagram.

    (________)(________) = ________________

    (___)(___) = ________________

    ___ = ___

34. Keep the grid showing. Use the x and y sliders to find the area in each table. There are many combinations of x and y for each area, so find as many different values as you can. Be prepared to verify your values of x and y using the method shown in step 33.

12

x	*y*
3	1
	2
0	
	0

16

x	*y*
	1
1	
	0

30

x	*y*

36

x	*y*

35. In the tables above, use values of x and y that are not integers. Use calculations to check the estimated values from the sketch. How many values can you find?

6

Geometry and Graphs in Grade 8

Mellow Yellow: Interpreting Graphs

ACTIVITY NOTES

INTRODUCE

Project the sketch for viewing by the class. Expect to spend about 5 minutes.

1. Open **Mellow Yellow.gsp** and go to page "Story 1."
2. Explain, ***Today you're going to put graphs into motion.*** Pointing to Mellow Yellow, explain, ***This is Mellow Yellow and she will be walking along this line segment, which represents the path from her house to the corner store. Of course, Mellow Yellow will not always walk in the same way. Sometimes she stops to pick something up or to rest. Sometimes she runs and sometimes she dillydallies.*** Press *Go!* ***Can anyone describe what Mellow Yellow did?*** Let students talk about the different things they noticed. Encourage them to talk especially about the different speeds at which Mellow Yellow moved. ***The graph provides a very nice way of describing Mellow Yellow's walk, and today you're going to work on being able to interpret the story a graph can tell.***
3. Show students how the points on the graph can be dragged to different positions and that the resulting walk done by Mellow Yellow will change. Try to avoid giving any of the segments a negative slope for now. Tell students that they will have an opportunity to change the graphs when they get to "Story 3," but that they should not change them for the first two stories.

DEVELOP

Expect students at computers to spend about 30 minutes.

4. Assign students to computers and tell them where to locate **Mellow Yellow.gsp.** Distribute the worksheet. Tell students to work through step 5 and do the Explore More question if they have time.
5. Let pairs work at their own pace. As you circulate, here are some things to notice.
   - Students often think that a slower speed means the graph goes down. Make sure students articulate what they see happening to Mellow Yellow's journey when the graph goes down, and how they can compare slower speeds to faster ones.
   - For worksheet step 3, where students are asked whether the graph corresponds to the story, invite students to think about how they could change the graph so that the story *does* match.
   - For worksheet step 4, encourage students to experiment with different locations of points 1 and 2 if they are having difficulty understanding the relationship of the graph to the movement of Mellow Yellow.

Invite students to change the locations of points 1 and 2, and describe a different story.

- For worksheet step 5, make sure students write out their stories fully. Ask them to connect each "leg" of the trip to the corresponding segment.

SUMMARIZE

Project the sketch. Expect to spend about 10 minutes.

6. Gather the class. Students should have their worksheets with them. Using page "Fit the Story," ask students to explain how they used the information in the story to know where to place the points on the graph.
7. Using page "Write a Story," ask two or three volunteers to describe their stories.
8. Ask students what kinds of motions they were able to create with the graph and also what kind of motions they could not create (such as acceleration).

EXTEND

1. Explain to students that acceleration and deceleration require nonlinear graphs. Point out that one of the problems with the graphs they used in Sketchpad is that Mellow Yellow would have a hard time starting immediately at a high speed. Instead she would probably start more slowly and accelerate until she achieved a high speed. On the board, draw an example of acceleration and ask students to use these curved lines to draw graphs that would better represent the stories about Mellow Yellow.
2. ***What other questions might you ask about graphing motion?*** Encourage all inquiry. Here are some ideas students might suggest.

   Why do steeper lines indicate faster speeds?

   Can you calculate the actual speed from the steepness of the line?

   What if a graphed segment of the trip were vertical?

ANSWERS

1. Answers will vary. Students should notice that the x-axis corresponds to the time it takes Mellow Yellow to travel and that the y-axis corresponds to the distance she has traveled. The y-coordinate of point *Stop* is just under 1 (mile) and its x-coordinate is just over 11, so it takes her just over 11 minutes to arrive at the corner store. There are three different slopes, including one that is 0 (corresponding to the horizontal segment) when Mellow Yellow is stopped.
2. In Story 2 Mellow Yellow goes backward and the corresponding segment slopes down.
3. The first segment should have a steeper slope than the last segment because she runs faster at the beginning. The middle segment should have a slope of 0 instead of a negative slope.
4. Answers will vary. Make sure that the first segment has a small positive slope, the second segment has a slope of 0, and the third segment is steeper than the first. Make sure also that Mellow Yellow runs far enough, so that the y-coordinate of *Stop* is just under 1.
5. Answers will vary. Make sure that the horizontal segments correspond to not moving and that the slopes of the other segments correspond appropriately to the speeds.
6. Answers will vary. Make sure that the y-coordinate of point *Stop* in both stories is just under 0.5 (miles) and that its x-coordinate is at 10 or 5 (minutes), depending on the story.

Mellow Yellow

Name:

In this activity you'll try to describe and predict how different motions, such as stopping, walking slowly, or walking very quickly, are represented on a graph.

EXPLORE

1. Open **Mellow Yellow.gsp** and go to page "Story 1." Press *Go!*, and then press *Show Story*. Describe how the features of the graph (the axes, slopes, and points) correspond to the story of Mellow Yellow's walk.

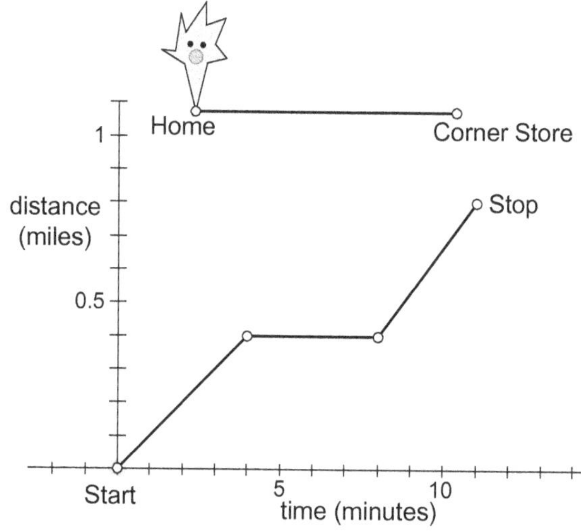

2. Go to page "Story 2." Read the story and compare the graph to the story. Then press *Go!*, and describe the different types of motion you see in Story 2 compared to Story 1.

3. Go to page "Story 3." Read the story and then press *Go!*. Decide whether the graph corresponds to the story. If not, change the graph (or the story!). Describe what you did.

4. Go to page "Fit the Story." Read the story and drag points 1 and 2 to make the graph fit the story. Check your graph by pressing *Go!*, and describe what you did.

5. Go to page "Write a Story." Write a story that fits the graph. Check your story by pressing *Go!*. Change your story if necessary. Then write your final story here.

EXPLORE MORE

6. Write your own story, but this time imagine that Mellow Yellow has to travel only to the bus stop, which is halfway to the corner store, in about 10 minutes. Then write a story in which she travels to the bus stop, but in about 5 minutes. Use pages "Explore More A" and "Explore More B" to fit the graph to each of your stories.

Shady Solutions: Graphing Inequalities on a Number Line

ACTIVITY NOTES

INTRODUCE

Project the sketch for viewing by the class. Expect to spend about 10 minutes.

1. Open **Shady Solutions.gsp.** Go to page "Number Line." Enlarge the document window so it fills most of the screen.
2. Drag point x back and forth on the number line. Ask, ***How would I graph the inequality $x \geq 2$?*** Students should respond that you need to shade the portion of the number line to the right of 2, including 2.
3. Explain what you're doing as you demonstrate. ***To shade the number line in this Sketchpad model, I position point x at 2, select the short vertical segment through x, and choose* Display | Trace Segment. *I use the arrow key to move the segment to the right and shade that part of the number line.*** Explain that all the shaded values, such as 2, 5, and π, satisfy the inequality $x \geq 2$, so they're part of the graph, whereas none of the unshaded numbers, such as 0, -3, and -10, satisfy the inequality, so they're not part of the graph. You may need to clarify the meaning of the word *satisfy* (to make the statement true). You might mention that when students make a graph like this on paper, they would put a filled-in dot at 2. Because the little segment in this Sketchpad model makes it easier to be precise, we are not using the dot.

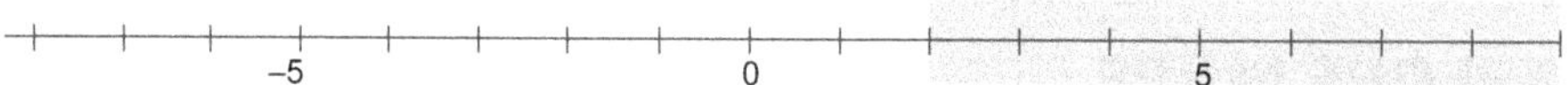

4. Go to page "A" and explain, ***In problem A there is a little segment that marks the expression x − 2 in addition to the segment that marks point x. Watch what x − 2 does while I drag x. Now we want to graph the slightly more complicated inequality x − 2 ≥ 5. We're still going to use the x-segment to shade values, but the relationship we're looking for is between x − 2 and 5. How should those markers be located? As I move x, tell me when the inequality is satisfied. You can all call out together. Start with x − 2 to the left of 5 and move x to the left.*** Students should call out for you to move the other way. Move right slowly. Students may start calling out when x reaches 5. Whether they do or not, build suspense at this point by saying, ***Is x − 2 greater than or equal to 5 yet?*** [No] Students should start calling out when $x - 2$ reaches 5.
5. Explain as you demonstrate. ***I start by moving x so that x − 2 lines up with 5. Then I select the x-segment and choose* Display | Trace**

Segment. ***Now, which way do I want to move x?*** [Right] ***I use the right arrow key to move x and shade that portion of the number line.***

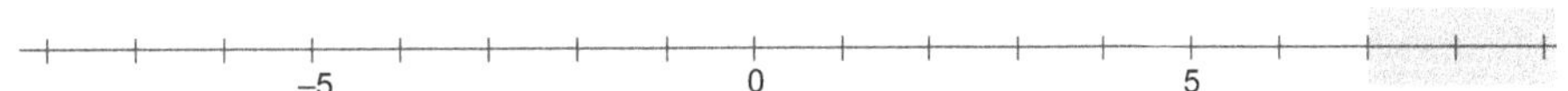

6. Explain, ***The shaded part represents all the solutions to the inequality*** $x - 2 \geq 5$. ***What simple inequality does this graph represent?*** [$x \geq 7$] ***So the inequality*** $x \geq 7$ ***is the solution to the inequality*** $x - 2 \geq 5$.
7. Tell students that they will graph the solutions to inequalities like this with increasingly complex expressions in them. ***Using the graphs you'll be able to write simple solution inequalities like x is greater than or equal to or x is less than or equal to some number. You'll record the graphs and the solutions on your worksheet. If you finish the pages with the given inequalities, go on to the "Explore More" and the "Make Your Own" pages.***

DEVELOP

Expect students at computers to spend about 25 minutes.

8. Assign students to computers and tell them where to locate **Shady Solutions.gsp.** Distribute the worksheet. Tell students to work through step 14 and do the Explore More if they have time. Encourage students to ask their neighbors for help if they are having difficulty using the sketch.
9. Let pairs work at their own pace. As you circulate, here are some things to notice.
   - Watch for students who might be tracing the wrong segment. Students might be tempted to trace a segment representing an expression instead of the *x*-segment. Remind them that the solution consists of the *x*-values that make the inequality true.
   - Make sure students are finding the correct solution graphs and inequalities and recording them on their worksheets. If you see a mistake, ask the student to go to that page and show you how he or she found the solution.
   - In worksheet step 12, students should observe that the solution to the inequality in problem C has its inequality symbol reversed from the original. Students' explanations of why this happens will vary. Refer them back to worksheet step 8 and encourage them to give as much detail as they can. They should observe that the expression

$-x + 2$ travels in the opposite direction from x because the expression includes the opposite of x.

- In worksheet step 13, students are asked to solve the inequalities algebraically. This activity is designed to introduce inequalities to students who have some experience with solving linear equations. For those students, as needed, suggest that they apply their equation-solving skills. You may need to make modifications, depending on the background of your students.

SUMMARIZE

Expect to spend about 10 minutes.

10. Gather the class. Students should have their worksheets with them. Begin the discussion by asking students to share their solutions to the inequalities.
11. Ask students if any strategies occurred to them for solving inequalities without Sketchpad. Whether or not they completed worksheet steps 13 and 14, students who compare their solutions to the original inequalities might note that the same strategies for solving equations (adding, subtracting, and dividing both sides of an equation) can also apply to inequalities.
12. Discuss worksheet step 12. Make sure students notice that the inequality symbol changed direction because of the $-x$. Depending on the background of your students, this observation can serve as a preview, introduction, or review of how to solve inequalities that involve multiplying or dividing by a negative number.
13. If time permits, discuss the Explore More and let students share inequalities they invented in the "Make Your Own" pages.

ANSWERS

4. Answers will vary. Students should observe that the marker for $x - 2$ moves in the same direction and speed as the marker for x and that the two markers are always two units apart.
5. $x \geq 7$

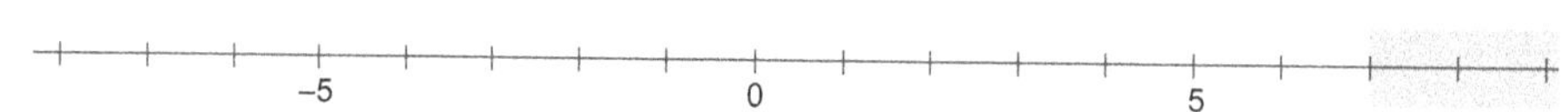

6. Answers will vary. Students should observe that the marker for $2x - 3$ moves faster than the marker for $x - 2$.

7. $x \geq 4$

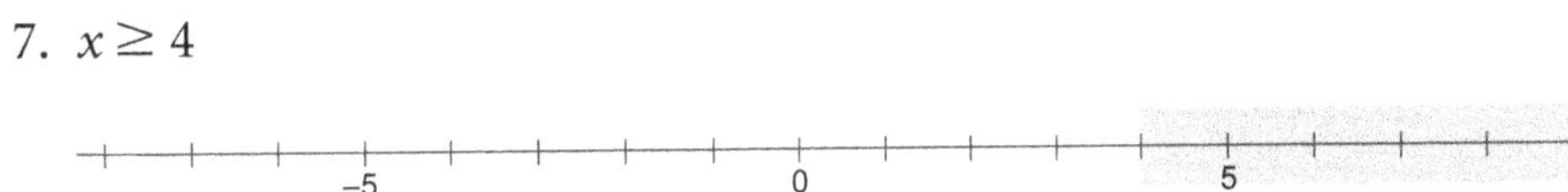

8. Answers will vary. Students should observe that the marker for $-x + 2$ moves in the opposite direction from the marker for x.

9. $x \leq -3$

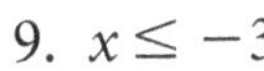

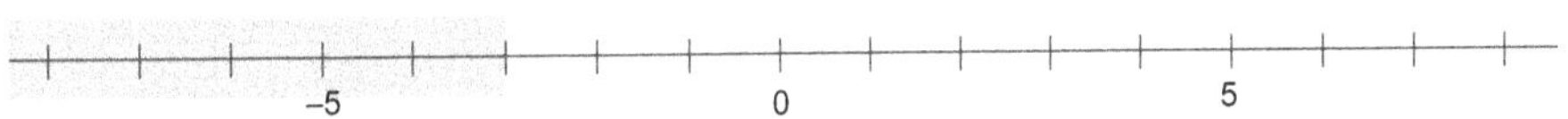

10. Answers will vary. Students should observe that the both the red and green markers move as x moves; on previous pages the green marker did not move, because it represented a constant value.

11. $x \leq 5$

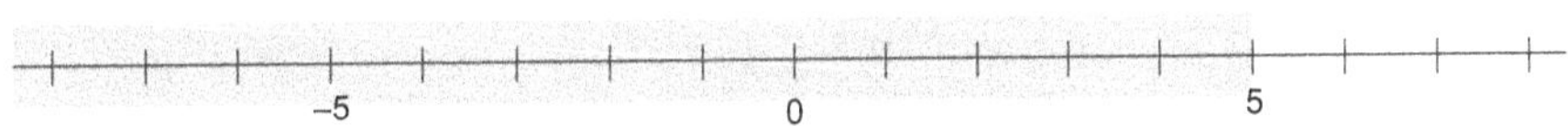

12. Problem C. The symbol in the original inequality, $-x + 2 \geq 5$, is switched in the solution, $x \leq -3$. Explanations will vary. Students should observe that the marker for the expression $-x + 2$ moves in the opposite direction as the marker for x. They may explain that this happens because x and $-x$ are opposites of one another.

13. A. $x \geq 7$
    B. $x \geq 4$
    C. $x \leq -3$
    D. $x \leq 5$

14. The solutions should match. For inequalities like problem C, you must reverse the inequality when multiplying (or dividing) both sides by a negative number.

15. $-1 \leq x \leq 5$

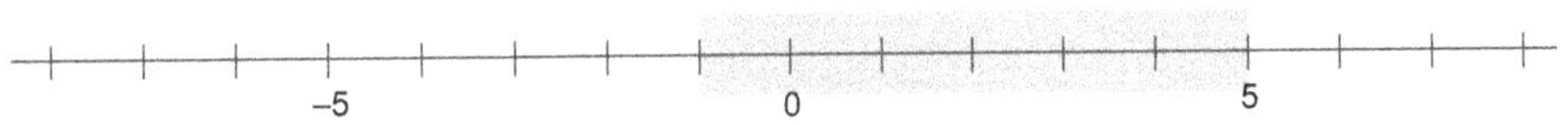

To solve this algebraically, split the compound inequality into two separate inequalities, $-5 \leq 2x - 3$ and $2x - 3 \leq 7$, solve them, and combine their solutions.

16. Answers will vary. Sample answer: $-2x + 1 \geq 9$ (solution is $x \leq -4$)

17. Answers will vary. Sample answers: $2x + 1 \leq 2x$ (no solutions), $2x + 1 \geq 2x$ (entire number line)

Shady Solutions

Name:

An inequality can have many solutions. You can represent them by shading the part of the number line that includes all the numbers that make the inequality true.

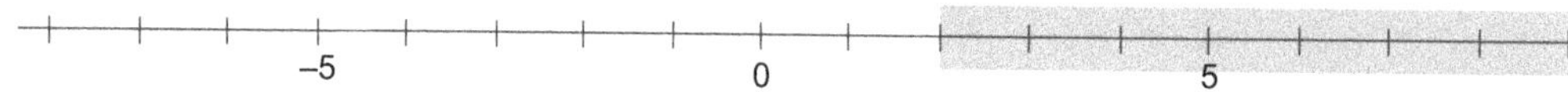

For example, the graph above shows the inequality $x \geq 2$. The shaded part includes all the numbers that make the inequality true.

What about a more complicated inequality, like $3x - 5 \leq 2x$? What values of x make it true? In this activity you'll explore graphs of solutions to inequalities like these and compare them to algebraic methods for finding these solutions.

EXPLORE

1. Open **Shady Solutions.gsp** and go to page "Number Line." Drag point x so that the inequality $x \geq 2$ is true.

2. Select the blue vertical segment through x and choose **Display | Trace Segment.**

3. Using the arrow keys to move x, shade the values of x on the number line that make the inequality $x \geq 2$ true. Your graph should end up looking like the example above.

4. Go to page "A." Drag x and observe how $x - 2$ behaves. How would you describe this behavior?

5. Trace the vertical segment through x and use the arrow keys to shade the solution to $x - 2 \geq 5$. Record the graph of the solution on the number line here.

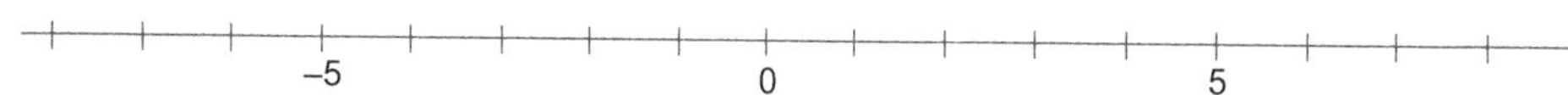

Write the solution by circling the correct inequality symbol and filling in the blank with a number.

$x \leq \geq$ ________

6. Go to page "B." Drag x and observe how $2x - 3$ behaves. How does this behavior differ from that of $x - 2$ on page "A"?

7. Repeat the process described in step 5 to shade the solution to $2x - 3 \geq 5$. Record your graph and solution here.

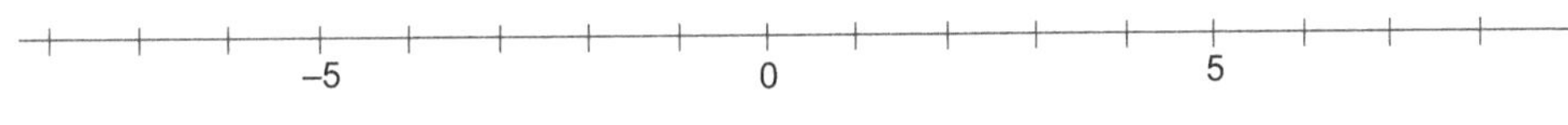

$x \leq \geq$ ________

8. Go to page "C." Drag x and observe how $-x + 2$ behaves. How does this behavior differ from those of pages "A" and "B"?

9. Find the solution to $-x + 2 \geq 5$. Record your graph and solution here.

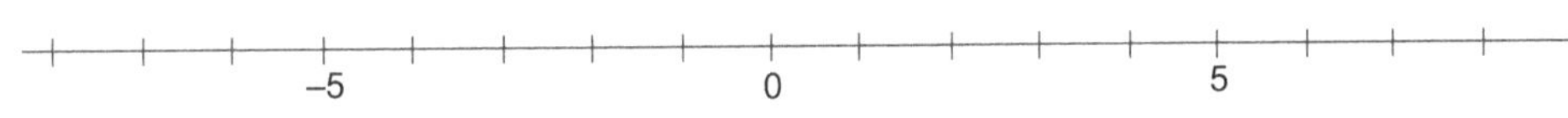

$x \leq \geq$ ________

10. Go to page "D." Drag x and observe what happens. How does this behavior differ from those of the pages "A" through "C"?

11. Find the solution to $3x - 5 \leq 2x$. This time the variable apprears on both sides of the inequality. Record your graph and solution here.

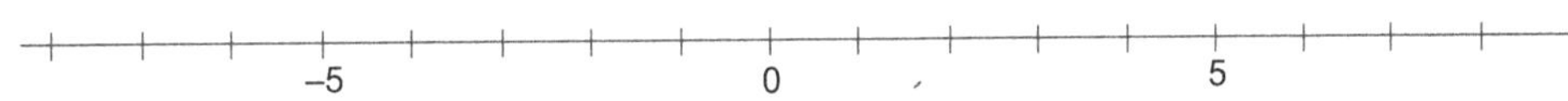

$x \leq \geq$ ________

12. In one of these four problems, the inequality symbol in the solution reversed direction. In which inequality did this happen? How is this related to the behavior of that expression?

13. Use algebraic techniques to solve each of these inequalities.

 A. $x - 2 \geq 5$

 B. $2x - 3 \geq 5$

 C. $-x + 2 \geq 5$

 D. $3x - 5 \leq 2x$

14. How do your solutions to step 13 compare to those in steps 5, 7, 9, and 11? What additional rule might you need to solve some inequalities?

EXPLORE MORE

15. Go to page "Explore More." Use the sketch to find the solution to the compound inequality $-5 \leq 2x - 3 \leq 7$. Then explain how you could solve this problem using algebraic techniques.

16. Go to page "Make Your Own." Double-click the red expression, green value, and both question marks to change them. Record the inequalities you invent, their solutions, and the graphs of the solutions. Try to invent an inequality in which the inequality symbol of the solution is reversed from the original.

17. Go to page "Make Your Own 2." Double-click both expressions to invent inequalities in which the variable appears on both sides of the inequality symbol. Record the inequalities you invent, their solutions, and the graphs of the solutions. See whether you can write an inequality that has no solutions or whose solution is the entire number line.

Amazing Angles: Finding Transversal Angle Pairs

INTRODUCE

Project the sketch for viewing by the class. Expect to spend about 5 minutes.

1. Open **Amazing Angles.gsp** and go to page "Maze."
2. Explain, ***Today you'll be looking at the special angles that are formed when a transversal line intersects a pair of parallel lines. There are a surprising number of related angles that are formed, as you can probably see just by looking at the sketch. There are so many, in fact, that sometimes it's hard to keep them straight! You'll get a chance to identify all the different pairs of special angles as you work on different a-maze-ing challenges. Before you begin, I'll demonstrate how the sketch works.***
3. Ask students to identify which angle is congruent to $\angle BAC$ and help them choose the correct button (either *Show Vertical Angle, Show Corresponding Angle,* or *Show Alternate Exterior*) to show that angle. Depending on which button has been chosen, a new angle will appear on the sketch. Ask students, ***Now how could we easily return to our original*** $\angle BAC$***?*** Students should be able to see that you can simply click the same button again so that if $\angle BAC$ is corresponding to $\angle BAD$, then $\angle BAD$ will be corresponding to $\angle BAC$. Make sure to try all three angles that are congruent to $\angle BAC$.
4. Ask students, ***Now which angles are not congruent to*** $\angle BAC$***?*** Students might point to the appropriate angle, but encourage them to figure out which button you would have to press (*Show Supplementary Angle*) to express the correct relationship. Point out that the *Supplementary Angle* button shows the supplementary angle along one of the parallel lines, but there's also a supplementary angle along the transversal.

DEVELOP

Expect students at computers to spend about 15 minutes.

5. Assign students to computers and tell them where to locate **Amazing Angles.gsp.** Distribute the worksheet. Tell students to work through step 6 and do the Explore More if they have time.
6. Let pairs work at their own pace. As you circulate, here are some things to notice.
   - Help students use the correct mathematical notation for describing angles (use $\angle BAC$ instead of $\angle A$).
   - If students are having trouble distinguishing between alternate interior and alternate exterior angles, use the *Show Interior* and *Show*

Exterior buttons, which shade the regions between the two parallels and outside the two parallels, respectively.

- For worksheet steps 2 and 3, the last cell in the second column does not need to be filled out. Cells in the second column contain the button students pressed to get from the adjacent cell to the first cell in the following row.

SUMMARIZE

Project the sketch. Expect to spend about 10 minutes.

7. Bring students together and discuss their results for worksheet steps 2 and 3. Ask students to list the different paths they found. Challenge students to come up with paths that use the least and the most number of different buttons.
8. Drag point *E* so that the transversal is slanted in the opposite direction. Ask students how their answers on the worksheet would change using this configuration.
9. Drag point *D* so that the parallel lines are no longer horizontal. Ask students how their answers on the worksheet would change using this configuration.
10. Challenge students to figure out how they could change the configuration of lines (not shown on the sketch) so that the relationships between pairs of angles break down.

EXTEND

1. Challenge students to add a transversal that would be parallel to transversal $\overleftrightarrow{GB}$ and to identify all the pairs of related angles. Challenge students to add another line parallel to the existing pair of parallel lines and to identify the pairs of related angles.
2. ***What other questions might you ask about angles?*** Encourage all inquiry. Here are some ideas students might suggest.

   What are names for noncorresponding angles on the same side of the transversal? How are they related?

   How do you know that these angles that you say are congruent are actually congruent?

   How many ways are there of visiting all 8 angles without repetition?

ANSWERS

1. Answers will vary.
2. Again, answers will vary. Here's one possibility: $\angle BAC$ (vertical angle), $\angle BAD$ (supplementary angle), $\angle EAC$ (vertical angle), $\angle GEF$ (corresponding angle), $\angle HEA$ vertical angle), $\angle AEF$ (supplementary angle), $\angle GEH$ (vertical angle)
3. Answer will vary. Here's one possibility: $\angle AEF$ to $\angle HEA$, to $\angle EAC$, to $\angle BAD$
4. Answer will vary. Here's one possibility: $\angle AEH$ (corresponding) to $\angle AEF$ (supplementary)
5. Answers will vary. Here's one possibility: $\angle DAE$ to $\angle EAC$ (supplementary), to $\angle AEH$ (alternate interior), to $\angle AEF$ (supplementary)
6. Answer will vary.
7. Answers will vary. $\angle DAB$ is congruent to $\angle HEA$ when the lines *DC* and *HF* are parallel. $\angle DAB$ is congruent to $\angle AEF$ when the lines *DC* and *HF* are parallel and both are perpendicular to line *BG* (in which case they are both right angles), or when the two blue lines intersect to form the congruent sides of an isosceles triangle.

Amazing Angles

Name:

In this activity you'll be solving several problems that will require you to find relationships between the eight angles formed at the intersections of two parallel lines with a transversal.

EXPLORE

1. Open **Amazing Angles.gsp** and go to page "Maze." Currently, $\angle BAC$ is marked. Use the buttons to visit every one of the eight angles created by the transversal line. Record the angles you've visited as you go. You may visit some angles more than once.

2. Start at $\angle DAE$ (you may have to click a few buttons to get there). Visit every one of the eight angles *only once.* Record the order of the angles and the buttons you pressed.

Angle	Button
$\angle DAE$	

3. Start at $\angle AEF$. How can you get to $\angle BAD$ in exactly three steps?

Angle	Button
$\angle AEF$	
$\angle BAD$	

4. Start at ∠*BAD.* Can you get to ∠*AEF* without visiting ∠*BAC* or ∠*CAE*? If so, record your steps.

5. Describe how you can visit all the interior angles without stepping in the exterior region. Start at ∠*DAE.* Try to find more than one way.

6. Create your own "angle maze" problem and write a solution.

EXPLORE MORE

7. Go to page "Explore More." Here there is a red transversal that cuts two nonparallel lines. Predict which pairs of angles are either congruent or supplementary. Drag the points on the sketch to change the relationship between lines *DC* and *HF* and drag points *A* and *E* to change the slant of the transversal. Explain the conditions under which ∠*DAB* is congruent to ∠*HEA.* Can ∠*DAB* ever be congruent to ∠*AEF*?

INTRODUCE

Project the sketch for viewing by the class. Expect to spend about 5 minutes.

1. Open **Menagerie.gsp.** Enlarge the document window so it fills most of the screen.
2. Go to page "Shapes." Explain, ***Here's a pentagon, which I can change by dragging any of its vertices. Today you'll look at a wide variety of transformations of this pentagon ABCDE. Before we start, can anyone think of a particular transformation that you've encountered before?*** Students might mention some of the more common congruence-preserving transformations such as rotation, reflection, and translation. After these three have been listed, ask, ***Does anyone know whether there are other kinds of transformations? If yes, how many do you think there might be?*** Students might think of dilation, or perhaps shearing. ***You're going to use Sketchpad to investigate some new transformations that you may never have seen before. You're going to focus especially on how each of these transformations changes ABCDE in specific ways. Before you begin, I'll demonstrate how the sketch works.***
3. Press *Show Menagerie.* Show students how they can drag the vertices of *ABCDE* to make all the shapes on the screen change. You might want to refer to these shapes using their labels or their unique colors. Press *Hide Menagerie.* Drag *ABCDE* so that it forms a recognizable shape such as a square. ***Can anyone guess what some of the other shapes will look like?*** Give students a chance to sketch out a few suggestions and ask for examples. Tell students, ***This may seem like a difficult task now, but it'll become much easier as you explore these shapes.***

DEVELOP

Expect students at computers to spend about 15 minutes.

4. Assign students to computers and tell them where to locate **Menagerie.gsp.** Distribute the worksheet. Tell students to work through step 8 and do the Explore More if they have time.
5. Let pairs work at their own pace. As you circulate, here are some things to notice.
   - Encourage students to drag *ABCDE* as they work through the worksheet. This will help them develop more general patterns and relationships.

- For worksheet step 3, students might have a difficult time finding any relationships. You can help them by directing their attention to geometric properties such as length, angle, area, and size. It may also help students to drag *ABCDE* into more recognizable shapes.

- For worksheet step 4, at first students may think that *R* is a rotation of *ABCDE*. Encourage them to keep changing *ABCDE* and watch how *R* changes. Use the word *flipped* or describe *R* as facing the other way. These are both good descriptions. It might be helpful also for students to realize that both *G* and *P* can be shifted along the plane to lie on top of *ABCDE*, but that *R* has to lift out of the plane and reflect to get back to *ABCDE*.

- For worksheet steps 5 through 7, encourage students to look for examples in one of the shapes. This will not be possible for worksheet step 6, although students may be able to think of a sheared shape that has the same side lengths as *ABCDE* with different angles. As needed, advise students to measure areas in worksheet step 7.

- For worksheet step 8, students can use a mix of formal and informal words to describe the transformations.

SUMMARIZE

Project the sketch. Expect to spend about 10 minutes.

6. Gather the class. Students should have their worksheets with them. Begin the discussion by reviewing worksheet step 3. Ask for volunteers to describe how the shapes resulting from transformations that don't preserve congruence are related to *ABCDE*.

7. Hide the shapes that are not congruent to *ABCDE* (shapes *M*, *T*, and *O*). Ask students to describe each visible shape and how it was transformed from *ABCDE*. List the properties each shape has in common with *ABCDE* and the properties it does not share.

8. If your students are familiar with the Transform menu commands, invite them to construct shapes on the projected sketch that result from the same types of transformations as the shapes already in the sketch, but that occupy different locations.

9. If time permits, discuss the Explore More. The vertices of shape *X* are the intersections of lines formed through this complex series of constructions: a line through the pre-image vertex and the center of a circle; the perpendicular to this line through the center; a line through

the pre-image vertex and the intersection of the perpendicular with the circle; a mirror line through the intersections of two lines with the circle; and a reflection of a line across the mirror line. The *Hint* buttons show this sequence for point *A*.

EXTEND

1. Invite students to describe another transformation, different from those in the shapes. You may challenge them to come up with another transformation that preserves congruence (the only other one would be a glide reflection) or other transformations that preserve only, for example, parallel lines (like a sheer transformation).
2. ***What other questions might you ask about transformations?*** Encourage all inquiry. Here are some ideas students might suggest.

   Don't congruent figures look alike? Are reflections really congruent?

   Is there a way, other than eyeballing, to know whether one shape is a transformation of another and what kind of transformation it is?

   How many different kinds of transformations are there?

ANSWERS

2. *R, P, G*
3. Answers may vary. *O* is a dilation of *ABCDE* by a factor of two. *T* has a horizontal dilation but no vertical dilation (it's a stretched version of *ABCDE*). *M* has a horizontal dilation of 0.5, but a vertical dilation of 2 (it's stretched vertically but squished horizontally).
4. *R* has a different orientation from the other shapes. By dragging *ABCDE* into a rectangle or square, *R* would no longer appear to have a different orientation.
5. Yes, *O* has the same angles, but the side lengths are twice as long.
6. No
7. Yes. *M* has the same area as *ABCDE*, but the two shapes are not congruent.
8. *P* is a rotation. *G* is a translation. *R* is a reflection. *O* is a dilation. *T* and *M* are both transformations that do not preserve angle, length, or shape.

Menagerie

Name:

In this activity you'll explore several different transformations of a pentagon.

EXPLORE

1. Open **Menagerie.gsp** and go to page "Shapes." Drag the vertices of pentagon *ABCDE.* Press *Show Menagerie.*

2. Drag the vertices of *ABCDE* again and observe what happens to each of the shapes. List the shapes that are congruent to *ABCDE.*

3. For each of the shapes that you did not list in worksheet step 2, describe the features it has in common with *ABCDE.*

4. Shapes *P* and *G* are congruent to *ABCDE.* All three are congruent to shape *R* too, but *R* has a distinguishing feature. Explain how it's different. How could you change the shape of *ABCDE* so that *R* would not seem to have this distinguishing feature?

5. Do any of these shapes have the same angle measures as *ABCDE,* but different side lengths? If so, which ones?

6. Do any of these shapes have the same side lengths as *ABCDE,* but different angle measures? If so, which ones?

7. Do any of these shapes have the same area as *ABCDE* without being congruent? If so, which ones?

8. Use what you know about each shape to figure out what type of transformation relates it to *ABCDE*.

EXPLORE MORE

9. Go to page "Explore More." This time *ABCDE* has been transformed in quite a different way. Drag the vertices of the pentagon to see how shape *X* changes. What relationships can you find between *X* and *ABCDE*? Look carefully at length, angle, and size. Compare your findings with the shapes you investigated on the previous page.

10. Press *Show Objects.* Describe what new relationships you see.

Transformers: Exploring Coordinate Transformations

ACTIVITY NOTES

INTRODUCE

Project the sketch for viewing by the class. Expect to spend about 5 minutes.

1. Open **Transformers.gsp** and go to page "Across." Enlarge the document window so it fills most of the screen.
2. Explain, ***You have learned about the different properties related to rotation, reflection, and translation. You might even find it easy to tell transformations apart just by looking at shapes and their images. But some applications of transformations, such as video game graphics, require comparing or creating transformations of shapes that you can't actually see. Coordinate geometry provides a very powerful tool to do this.***
3. Drag the vertices of *ABCD* and say, ***Here's a quadrilateral that I can move and change by dragging its vertices. Now I'm going to apply a mysterious transformation to it by using one of my Transformers.*** Model how to use **Transformer #1** by choosing it from the Custom Tools menu and clicking the vertices of *ABCD* in order. ***Which transformation produced this image of ABCD?*** [Reflection across the *x*-axis] Then show students how the labels of the two quadrilaterals can help them figure out which are the corresponding coordinates.
4. Select point *A* and model how to measure its coordinates. Then select point *A1* and ask, ***Can you predict the relationship between the coordinates of point A and point A1? This is the type of relationship you'll be looking for today. Understanding how the coordinates of the original shape are related to those of its image is very helpful because it allows you to transform a shape without finding a line of reflection or a center of rotation.***

DEVELOP

Expect students at computers to spend about 30 minutes.

5. Assign students to computers. Tell them where to locate **Transformers.gsp.** Distribute the worksheet. Tell students to work through step 9. Encourage students to ask their neighbors for help if they are having difficulty with the custom tools.
6. Let pairs work at their own pace. As you circulate, here are some things to notice.
   - Help students use the custom tools correctly by clicking on the vertices of *ABCD in order* around the shape.

- Make sure students choose the **Arrow** tool and drag the vertices of *ABCD* after using a custom tool.
- If students have difficulty seeing the relationship in worksheet step 4, suggest that that they keep *ABCD* in Quadrant II.
- Help students organize the coordinate measures so that they can more easily compare corresponding coordinates.
- Show students that the custom tools can be used on any quadrilateral, not just *ABCD*.

7. When most students have reached worksheet step 9, gather the class together and make sure that they have been able to describe the coordinate relationships in worksheet steps 5 and 8. If students have not done so, you might introduce the notation $(x, y) \rightarrow (new\ x, new\ y)$ to describe these relationships. Explain, ***You have looked at two types of reflection that changed either the x-coordinate or the y-coordinate. Can you predict what might happen with other transformations?*** Then ask students to work through step 20 and do the Explore More if they have time.

SUMMARIZE

Project the sketch. Expect to spend about 10 minutes.

8. Ask students to describe how each transformer works. Help them describe the transformations in terms of arbitrary coordinates. For example, **Transformer #1** takes a point (x, y) and reflects it to a point $(x, -y)$. **Transformer #2** takes a point (x, y) to $(-x, y)$. Mention that these descriptions apply only to very specific transformations (reflections over the axes and rotations around the origin), and that reflections across other lines, and rotations around other points, may not be so easily described by coordinates.
9. To summarize the transformations, go to page "Match." Challenge students to describe the Transformers (and transformations) that would transform the quadrilateral to each of the three images. Students may want you to measure the coordinates of the three holes. One of the transformations is an x-shift of 3 and a y-shift of -4; another is a rotation by 270° counterclockwise (or 90° clockwise); and the third is a reflection across the line $y = x$ (which would be easier for students to see if you first assigned Extension 1). Many students may suggest that this last reflection is a rotation by 180°. Use the relationships between

the coordinates they discovered to help them understand why it is not a rotation.

10. If time permits, discuss the Explore More. Invite students to share their observations about sequences of transformations.

EXTEND

1. Have students construct a line that passes through the origin with a slope of 1 (with equation $y = x$). Tell them to mark this line as a mirror and then use **Transform | Reflect** to reflect *ABCD* across this line. Ask them to explore the relationships between the corresponding coordinates. Later, change the slope of the line so that students can see that this relationship will break down.
2. ***What other questions might you ask about transformations?*** Encourage all curiosity. Here are some ideas students might suggest.

   Will it always be true that using two transformations in a row is the same as a single transformation?

   What's going on when I choose a Transformer and don't follow the vertices in order?

   What kind of transformation just switches the two coordinates?

ANSWERS

4. The image is a reflection of *ABCD* across the x-axis.
5. The x-coordinates are the same and the y-coordinates have the opposite sign. Using symbols, $(x, y) \rightarrow (x, -y)$.
7. The image is a reflection of *ABCD* across the y-axis.
8. The x-coordinates have the opposite sign and the y-coordinates are the same. Using symbols, $(x, y) \rightarrow (-x, y)$.
9. Use **Transformer #1** on the quadrilateral in Quadrant I or **Transformer #2** on the quadrilateral in Quadrant III.
11. The image is a translation of *ABCD.*
12. The y-coordinates are the same and the x-coordinates of the image are 8 greater than those of *ABCD.* Using symbols, $(x, y) \rightarrow (x + 8, y)$.

13. To land the image on the target, use an x-shift of 10.0 and a y-shift of -6.0.

14. Answers will vary, but both the x- and y-coordinates should be negative.

16. The image is a rotation of *ABCD* by 90° counterclockwise around the origin.

17. The x-coordinate of each point in *ABCD* becomes the y-coordinate of the corresponding point in the image. The y-coordinate in *ABCD* becomes the opposite of the x-coordinate in the image. Using symbols, $(x, y) \to (-y, x)$.

19. The image is a rotation of *ABCD* by 180° around the origin. The x-coordinate of each point in *ABCD* becomes the opposite of the x-coordinate in the image. The y-coordinate becomes the opposite of the y-coordinate in the image. Using symbols, $(x, y) \to (-x, -y)$.

20. Answers will vary. Sample solutions: Use **Transformer #2** on *ABCD.* Use **Transformer #1** or **Transformer #4** on the image in Quadrant IV. Use **Transformer #5** on the image in Quadrant III. Other combinations are possible.

21. Answers will vary. The example given will produce a rotation of *ABCD* by 180° around the origin. This can also be accomplished by using the two Transformers in the reverse order, or by using **Transformer #5.**

Transformers

Name:

In this activity you'll use transformers—tools that transform a shape into a new shape—to explore the relationship between the coordinates of the original and transformed shapes.

EXPLORE

1. Open **Transformers.gsp** and go to page "Across." Drag the vertices of quadrilateral *ABCD* into any shape you like.

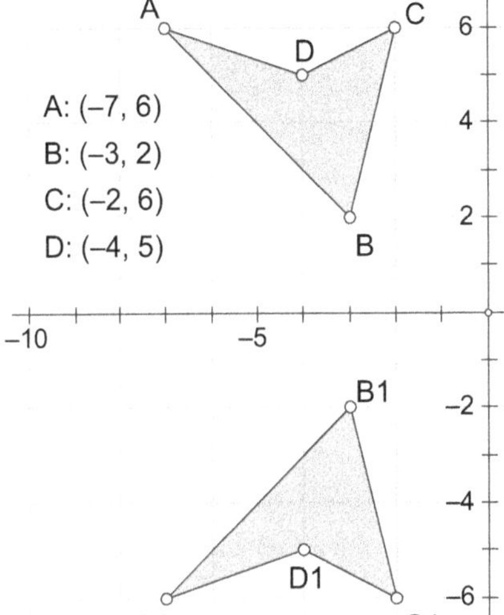

2. Select the vertices of *ABCD* and choose **Measure | Coordinates.**

3. Choose **Transformer #1** from the Custom Tools menu. Click each vertex of *ABCD* in a clockwise (or counterclockwise) order. You should see a new quadrilateral, called the *image,* in Quadrant III.

4. Drag the vertices of *ABCD* or the quadrilateral interior. Describe how the two quadrilaterals are related.

5. Measure the coordinates of the image. Describe the relationship between the corresponding coordinates of the two quadrilaterals.

6. Use the custom tool **Transformer #2** on *ABCD.*

7. Drag the vertices of *ABCD.* Describe how the new image is related to *ABCD.*

8. Measure the coordinates of the new image. Describe the relationships between the corresponding coordinates of *ABCD* and the new image.

9. Use either **Transformer #1** or **Transformer #2** to create a quadrilateral in Quadrant IV. Find two different ways of doing this.

10. Go to page "Along." Use **Transformer #3** on *ABCD.*

11. Drag the vertices of *ABCD.* Describe how the two quadrilaterals are related.

12. Measure the coordinates of *ABCD* and the image. Describe the relationship between their corresponding coordinates.

13. Experiment with changing the *x*-shift and *y*-shift values. To change a value, either double-click it and enter a new value, or select it and use the + and − keys on your keyboard. Then press *Show Target.* What shift values will make the image land on the target?

14. Drag *ABCD* so that it is in Quadrant I only. What shift values will place the image in Quadrant III?

15. Go to page "Around." Use **Transformer #4** on *ABCD.*

16. Drag the vertices of *ABCD.* Describe how the two quadrilaterals are related.

17. Measure the coordinates of *ABCD* and the image. Describe the relationship between their corresponding coordinates.

18. Use **Transformer #5** on *ABCD.*

19. Measure the coordinates of the new image. Describe how the new image is related to *ABCD* and the relationship between their corresponding coordinates.

20. Find three different ways of producing another image in Quadrant I using any of the Transformers.

EXPLORE MORE

21. Go to page "Explore More." Experiment with different sequences of five Transformers. What happens if you first use **Transformer #1** and then use **Transformer #2** on the image? What other Transformers or sequences of Transformers will accomplish the same result?

Rainbow of Lines: Investigating Slope and Intercepts

INTRODUCE

Project the sketch for viewing by the class. Expect to spend about 5 minutes.

1. Open **Rainbow of Lines.gsp.** If your students are not familiar with function notation, you might start with page "Function," which provides a brief demonstration of function notation. For this activity students simply need to know that the equations of lines will be displayed as functions rather than the more typical $y = \ldots$ form.

2. Go to page "Lines." Explain, ***Today you'll work on arranging these lines into different patterns using color, slope, and intercepts. Before you begin, I'll demonstrate how the sketch works. As you can see, the red line has been matched with function f. What is the slope of this line?*** Students should be able to respond that the slope is $\frac{1}{3}$. Encourage students to explain how they know the slope is $\frac{1}{3}$, using both algebraic evidence (looking at the value of the coefficient of x in the function) and graphical evidence (looking at the fact that the slope is positive, but less than 1). ***You can also see that the blue line is matched with function r. What is the slope of the blue line?*** Again, ask for algebraic and graphical evidence.

3. ***How could we change the equation of the blue line so that it went through the same quadrants as the red line?*** Students may have different ideas, including simply making the slope $+5$ instead of -5. Show students how they can edit the definition of the blue function to change the graph. ***All the lines go through the origin. How could we change the blue line so that it did not pass through the origin?*** Give students an opportunity to try different ideas. Some students may want to change the slope again, whereas others may think of adding a constant term to the function. Allow them to experiment with several different options. Then, tell students, ***Now you'll match the remaining functions to their corresponding graphs.***

4. If you want students to save their work, demonstrate choosing **File | Save As,** and let them know how to name and where to save their files.

DEVELOP

Expect students at computers to spend about 20 minutes.

5. Assign students to computers and tell them where to locate **Rainbow of Lines.gsp.** Distribute the worksheet. Tell students to work through step 6 and do the Explore More if they have time.

6. Let pairs work at their own pace. As you circulate, here are some things to notice.
   - Make sure students are able to edit the equations.
   - Help students recognize that lines going through the origin have equations of the form $y = mx$.
   - Help students notice that parallel lines have the same slope.
   - In worksheet steps 4 and 5, students may have difficulty finding x-intercepts of graphs. You may have to provide several examples to help them see that the value of the x-intercept is $\frac{-b}{m}$. If appropriate, show students how this value comes from setting the equation equal to zero.

SUMMARIZE

Project the sketch. Expect to spend about 5 minutes.

7. Gather the class. Students should have their worksheets with them. Use the sketch to support the discussion. Review worksheet steps 2 and 3 to make sure students understand the slope-intercept form of a line: $y = mx + b$. If needed, you can choose the custom tool **Slope Triangle** and click on a line to create an adjustable slope triangle.
8. Review worksheet step 4. Ask students to generate examples of equations whose graphs have an x-intercept of -3. Change the existing graphs to illustrate their examples.
9. Using one of the graphs already on the sketch, ask students to work quickly in pairs to generate another line parallel to it. Create as many of these as you can on the sketch, assigning different colors if possible. Ask students what you would need to do to make all the lines slant in the opposite way, or to be horizontal, or to be vertical (which is not possible).

EXTEND

1. Choose **Number | New Parameter** and name it m. Change each equation so that the slope is defined by m. Select parameter m and choose **Display | Animate** to illustrate how all the lines change slope as the parameter changes.

2. ***What other questions might you ask about equations and lines?*** Encourage all inquiry. Here are some ideas students might suggest.

   Could you calculate the slope of a line on a graph without knowing its equation?

   Can you tell from the equations whether two graphs have the same x-intercept?

   What is the equation of a vertical line? Can you graph it as a function?

ANSWERS

2. All the slopes are positive (and all the y-intercepts are zero).
3. Answers will vary. Students should generate equations of the form $y = mx - 3$.
4. Answers will vary. Students should generate equations in which $\frac{-b}{m} = 2$.
5. It's the same line.
6. Answers will vary. All the lines should have the same slope, and the y-intercepts should be separated by a constant difference.
7. Answers will vary. One line will have a slope of 4 and the other a slope of 0.
8. Answers will vary depending on equations students choose in the previous step.

Rainbow of Lines

Name:

In this activity you'll investigate how to change equations (expressed in function form) in order to produce different linear graphs. Then you'll use color to create a rainbow of lines.

EXPLORE

1. Open **Rainbow of Lines.gsp** and go to page "Lines." The red and blue graphs are already matched to their equations. Use **Display | Color** to match the three black graphs to their corresponding equations.

2. To change an equation, double-click it and then edit the expression in the function calculator. Change the equations so that all the graphs pass through the origin and remain in Quadrants I and III. What can you say about their slopes?

3. Change the function rule for equation *f* to $x - 3$. The graph now goes through $(0, -3)$, which is the *y*-intercept of the line. Change all the equations so that their graphs all go through $(0, -3)$. Describe your strategy.

4. Change the function rule for equation *f* to $3x - 6$. The graph now goes through $(2, 0)$, which is the *x*-intercept of the line. Change the function rule for equation *g* to $-2x + 4$. The graph for equation *g* now has the same *x*-intercept as that of equation *f*. Change the remaining equations so that their graphs all have the same *x*-intercept. Describe your strategy.

5. Change equation *r* so that its graph has the same *x*-intercept and same slope as equation *f*. What do you notice?

6. Change the equations so that the graphs create a rainbow. The lines should have the same slope and be equally spaced. You may want to add more lines by choosing **Graph | Plot New Function** to create a rainbow with the full spectrum of colors. Describe your strategy.

EXPLORE MORE

7. Change the function rule for equation *f* to $4x - 4$ and the rule for equation *g* to 2. If the graphs of those two equations form two sides of a parallelogram, then change equations *h* and *p* so that their graphs complete the parallelogram.

8. Can you change equation *r* so that its graph is the diagonal of the parallelogram from the previous step?

Dance Pledges: Plotting Linear Equations

INTRODUCE

Project the sketch for viewing by the class. Expect to spend about 10 minutes.

1. Open **Dance Pledges Present.gsp,** and go to page "Problem." Enlarge the document window so it fills most of the screen. Ask for volunteers to read the problem aloud. You might have students act it out.

2. ***In a few minutes, you're going to work in pairs to compare the three ideas for getting pledges. But first, does their reasoning make sense?*** Allow students to share their opinions and discuss what they think is realistic.

3. Go to page "Plots" and lead students through plotting a function for Chris's scheme. ***How would you calculate how much Chris's scheme earns if Chris dances for 6 hours?*** [$9] ***8 hours?*** [$12] ***24 hours?*** [$36] ***What calculations are you doing?*** [$1.50 times the number of hours] Choose **Graph | Plot New Function** and enter the expression $1.5 * x$.

4. Demonstrate how to construct a point on the graph by selecting the graph and choosing **Construct | Point on Function Plot** or by using the **Point** tool. Then select the point and choose **Measure | Coordinates.** Help students focus on the meaning of the coordinates. ***Can we use the coordinates to find out how many hours Chris would have to dance to earn $18?*** [Drag the point until the y-coordinate is 18 and find the x-coordinate. He would have to dance 12 hours.]

DEVELOP

Expect students at computers to spend about 25 minutes.

5. Assign student pairs to computers and tell them where to locate **Dance Pledges.gsp.** Distribute the worksheet. Tell students to work through step 12 and do the Explore More if they have time.

6. Let pairs work at their own pace. As you circulate, here are some things to notice.

   - Allow plenty of opportunity for students to help each other. Use questions similar to those you asked in the introduction: ***How would you calculate how much Jesse's*** (or ***Dale's***) ***scheme earns for 6 hours?*** [$10 (or $6)] ***8 hours?*** [$10 (or $7)] ***What calculations are you doing?***

   - You might need to show students how to adjust the scales on the axes to see their plots.

- If students don't see the **Plot New Function** choice in the Graph menu, they have some other object selected. They can click in blank space to deselect all objects.
- Jesse's scheme might be challenging because students may believe that the function expression must contain an x.
- In Dale's plan, students might use 50 instead of 0.5 for the hourly amount. ***What measurement units are you using?***
- You might suggest that students label the graphs so that they know which represents whose plan.

SUMMARIZE

Continue to project the sketch. Expect to spend about 10 minutes.

7. Have a few selected pairs share their insights. Introduce or review the terms *slope* and *y-intercept*. ***Can you tell what the slope and y-intercept are from the equation?*** [Yes] ***If you know that the slope was 3 and the y-intercept was 7, what would the equation be?*** [$y = 3x + 7$]
8. Encourage discussion of the question in worksheet step 9. The answer depends on the assumptions made about dancing parts of hours. Students may argue that the payments are only by the hour, so Chris's scheme never pays exactly $10. Depending on the background of your students, you can also have them use substitution to find the points of intersection algebraically.
9. ***What have you learned?*** Bring out these objectives.
   - The graph of $y =$ (hourly amount) $x +$ (fixed amount) is a straight line.
   - The amount multiplied by x is the slope of the line, measuring its steepness.
   - The amount added is the y-intercept of the line, measuring its height along the y-axis.
   - The graph of $y =$ (fixed amount) is a horizontal line.
   - A point of intersection of two graphs can show where two different plans earn the same amount of money for the same amount of work.

10. ***What other questions can you ask?*** Some interesting questions are these.

    What if the slope or y-intercept were negative? What would the equation and graph look like?

    Are there any lines that can't be plotted?

ANSWERS

1. $y = 1.5x$
2. $30
3. $y = 10$
4. Answers will vary. Both graphs are straight lines. Jesse's graph is horizontal but Chris's is not. (It rises from left to right.)
5. $10
6. $y = 5 + 0.5x$ or $y = 0.5x + 5$
7. The earnings for dancing 0 hours.
8. $15
9. Jesse's scheme always brings in $10 per donor. Dale's scheme brings in $10 for dancing 10 hours. Chris's scheme brings in $10 for dancing $6\frac{2}{3}$ hours.
10. Points representing the same amount of money for the same number of hours are at the intersections of the graphs. Chris's and Dale's plans both earn $7.50 for 5 hours of dancing. Dale's and Jesse's plans both earn $10 for 10 hours of dancing.
11. As the hourly amount increases, the line gets steeper, or the graph rises more for each hour danced.
12. As the fixed amount increases, the height of the line as it passes through the y-axis increases.
13. Answers will vary.

Dance Pledges

Name:

Three friends are planning to be in a dance marathon to raise money for kids with cancer.

Chris says, "I'm going to dance for the whole 24 hours, but nobody thinks I'll make it more than 2 or 3 hours, so I'll ask them to pledge $1.50 an hour."

Jesse says, "I know I can't go for more than a few hours, so I'm going to ask for $10 no matter what."

Dale says, "I have no idea how long I can go. I think I'll ask for $5 plus 50 cents an hour."

1. Open the document **Dance Pledges.gsp** and go to page "Plots."

   Plot a function for Chris's plan, using **Graph | Plot New Function.** What equation did you use?

2. Construct a point on the graph. Select the point and choose **Measure | Coordinates.**

   Drag the point. How much will each donor pay if Chris dances 20 hours?

3. Plot a function for Jesse's plan. What equation did you use?

4. How does this graph compare to Chris's?

5. Construct a point on Jesse's graph and measure its coordinates.

   Drag the point. How much will each donor pay if Jesse dances 20 hours?

6. Plot a function for Dale's plan. What equation did you use?

7. What does the point where the graph intersects the y-axis represent?

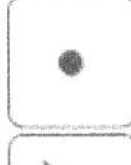

8. Construct a point on the graph and measure its coordinates.
   Drag the point. How much will each donor pay if Dale dances 20 hours?

9. For how much dancing time does each plan bring in $10 per donor? Explain.

10. Do any of the plans ever bring in the same amount of money for the same number of hours danced? Explain.

11. Go to page "Sliders." You'll see sliders for fixed and hourly amounts in a pledge plan. Choose **Graph | Plot New Function.** Instead of entering numbers, click *Fixed Amount* and *Hourly Amount*. What happens to the graph as you change the *Hourly Amount* slider? Explain.

12. What happens to the graph as you change the *Fixed Amount* slider? Explain.

EXPLORE MORE

13. Create a scheme that does not produce a straight line when graphed.

Hikers: Solving Through Multiple Representations

INTRODUCE

Project the sketch for viewing by the class. Expect to spend about 10 minutes.

1. Open **Hikers Present.gsp** and go to page "Problem." Enlarge the document window so it fills most of the screen. Read the problem aloud. Pause when you come to the first occurrence of mi/h, and ask students how to read it and what it means.
2. You might have two volunteers act out the problem for the class. Then go to page "Simulation" and press *Start/Stop Simulation.* ***Between what times do the hikers meet?*** [Between 3 and 4 hours].
3. ***What might we do to find the time more precisely?*** Encourage many suggestions. Go to page "Table".
4. ***Do the distances corresponding to time 0 make sense?*** Help students see that the distances are from the trailhead, not the distances traveled, so Maria begins at 12 miles.
5. ***What should go on the row corresponding to time 1?*** Elicit the idea that Edna will be 1.5 miles from the trailhead and Maria will be $12 - 2 = 10$ miles from the trailhead. Demonstrate how to double-click the parameters to change their values.

DEVELOP

Expect students at computers to spend about 25 minutes.

6. Assign student pairs to computers and show them how to find **Hikers.gsp.** Distribute the worksheet. Ask students to work through step 12 and do the Explore More if they have time.
7. Let pairs work at their own pace. As you circulate, here are some things to notice.
   - Some students may say that the hikers meet 6 miles from the trailhead, because 6 appears in both columns (or because 6 is the halfway point). Ask, ***How many hours after they left were they 6 miles from the trailhead? Can you say that's the same time? What does it mean to meet?***
   - Students may have difficulty reading the table to see when the hikers meet. ***Who was closer to the trailhead after 2 hours? After 3 hours? After 4 hours?***
   - In worksheet step 3, if students don't see **Graph | Plot as (x, y),** they probably still have the previous point selected. Tell them to click in blank space before they plot each point.

- If the lines that students trace in worksheet steps 7 and 8 do not pass through the plotted points from step 3, tell them to change their expressions in step 6, erase their traces, and try again.
- The coloring in worksheet step 9 is intended to help students associate the graphs directly with the table.
- In worksheet step 10, students may not realize that the point they seek is the intersection. Help them think of one variable at a time. ***What points represent the two hikers at a distance 9 miles from the trailhead? What are their times when they are there? What points represent the hikers 6 miles from the trailhead? What are their times? What point represents when they are at the same place at the same time?***
- As needed, help students realize in worksheet step 11 that to find exact times and distances, they must convert to fractions.

SUMMARIZE

Expect to spend about 10 minutes.

8. Reconvene the class. Select some pairs to present their sketches (or use those in **Hikers Present.gsp**). Discuss worksheet steps 5–11.
9. ***Which approach do you prefer: a table, a graph, or an equation?*** Student preferences will vary. Students should realize that numerical information can be represented in multiple ways: arithmetically, algebraically, and graphically. Encourage comparisons of the methods, such as the fact that the table and graph give only estimates, whereas the equation could give an exact answer.
10. ***What have you learned?*** You may wish to have students respond individually in writing to this prompt, or have volunteers respond verbally. Bring out these objectives.
    - The same situation can be represented with tables, graphs, or equations.
    - Making a table can help in finding expressions to graph.
    - Knowing expressions to graph can help in finding an equation to solve.

EXTEND

What other questions might we ask? Encourage all student curiosity. Mathematical questions of interest include these.

Why are the graphs straight lines?

Why does the point of intersection represent where the hikers met?

Is there an easier way to find an exact solution?

What if the hikers paused to rest? Could we still tell when they met?

Are there ways other than tables, graphs, and equations to represent the situation?

What if there were no Maria? If Edna went over one day and came back the next, would there necessarily be a point at which she was at the same place at the same time of day?

ANSWERS

1.

Time (hours)	Edna's Distance (miles from trailhead)	Maria's Distance (miles from trailhead)
0	0.0	12.0
1	1.5	10.0
2	3.0	8.0
3	4.5	6.0
4	6.0	4.0
5	7.5	2.0

2. The hikers will pass each other between hour 3 and hour 4. Any students who put in extra rows between hour 3 and hour 4 may predict a smaller interval.

3. The points representing each hiker's distances lie in a straight line, and those lines will cross, though not at a data point. The sketch should look like this.

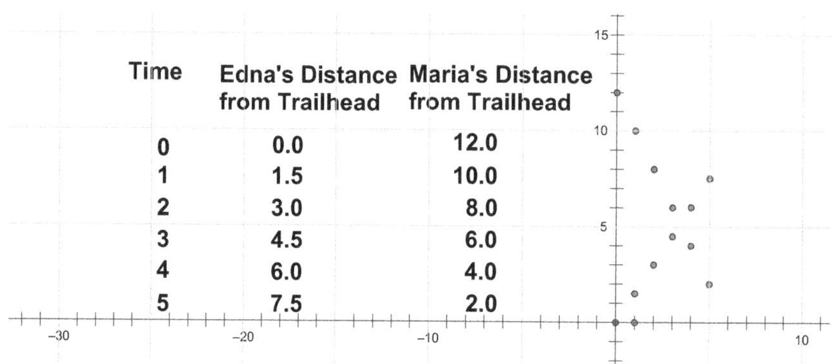

6. Edna: $1.5x$; Maria: $12 - 2x$
7. The traced point moves in a straight line, passing through the plotted points representing Edna's distance.
8. The traced lines intersect when *Time* is approximately 3.4 hours.
10. Answers may vary. The point of intersection will be approximately (3.43, 5.14), indicating a time of 3.43 hours and a distance of 5.14 miles from the trailhead.
11. $1.5x = 12 - 2x$

    $3.5x = 12$

    $$x = \frac{12}{3.5} = \frac{12}{\frac{7}{2}} = \frac{24}{7}\text{ hours}$$

    $$1.5\left(\frac{24}{7}\right) = \frac{3}{2} \cdot \frac{24}{7} = \frac{36}{7}\text{ miles}$$

12. $\frac{24}{7}$ hours $= 3\frac{3}{7}$ hours

    $= 3$ hours $+ \frac{3}{7} \cdot 60$ minutes

    $= 3$ hours and about 25 minutes

    Some students may go further to add about 43 seconds.

13. Answers will vary. Students familiar with solving systems of equations simultaneously may write two equations and solve by substitution (equivalent to what is done in worksheet step 8) or by elimination.

    $$\begin{aligned} 1.5x - y &= 0 \\ 2x + y &= 12 \\ \hline 3.5x \quad &= 12 \end{aligned}$$

    $$x = \frac{12}{3.5} \approx 3.43\text{ hours}$$

    $$y \approx 1.5(3.43) \approx 5.14\text{ miles}$$

    Alternatively, students might think about closing speed. The hikers together need to cover 12 miles at a combined speed of $1.5 + 2 = 3.5$ miles per hour. Doing so will take $\frac{12}{3.5}$ hours.

    Solutions could also be estimated by finding the intersection of the graphs of equations $y = 3.5x$ and $y = 12$.

Hikers

Name:

Edna leaves a trailhead at dawn to hike toward a lake 12 miles away, where her friend Maria has been camping. At the same time, Maria leaves the lake to hike toward the trailhead (on the same trail, but in the opposite direction).

Edna is walking uphill, so her average speed is 1.5 mi/h. Maria is walking downhill, so her average speed is 2 mi/h.

In this activity you'll investigate when and where the hikers will meet.

EXPLORE

1. Complete the table.

Time (hours)	Edna's Distance (miles from trailhead)	Maria's Distance (miles from trailhead)
0	0	12
1		
2		
3		
4		
5		

2. Open **Hikers.gsp** and go to page "Table." Double-click each value and change it to match the value in the table above. From the table, what can you predict about when the hikers will meet?

To predict the meeting time more accurately, you'll graph the points in the table, and then graph lines through them.

3. To graph Maria's first point, select the 0 in the Time column, then the 12 in Maria's Distance column, and choose **Graph | Plot as (x, y).** Axes and a grid will appear, with the point (0, 12) plotted. Repeat this process to plot all of Edna's and Maria's distances in the table. What patterns do you see?

4. Construct a point on the *x*-axis. While the point is still selected, choose **Measure | Abscissa (x).**

5. Change the label of the point on the axis to *T* and the label of the abscissa measurement to *Time.*

6. Choose **Number | Calculate** and use the Calculator to enter an expression for each hiker's distance from the trailhead. To enter *Time* into the Calculator, click the value in the sketch. What expressions did you use?

   Edna:

   Maria:

7. Select *Time* and your value for Edna's distance from step 6. Choose **Graph | Plot as (x, y)** and then choose **Display | Trace Plotted Point.** Drag point *T* and describe what you see.

8. Repeat step 7 using *Time* and your value for Maria's distance. When will the two hikers meet?

9. Choose **Display | Erase Traces.** Choose **Graph | Plot New Function** and enter an expression for each hiker's distance from the trailhead after *x* hours. Color each graph with the same color as the column it represents.

10. Construct a point on Maria's graph. While it's selected, choose **Measure | Coordinates.** Drag this point to predict when and where the hikers will meet. What are coordinates of that point? What does each coordinate represent?

11. Use the expressions you graphed to write a single equation. Solve the equation for the time when they meet, and then find the distance from the trailhead. Show your work.

12. Convert your time solution to hours and minutes.

EXPLORE MORE

13. You've made predictions by a table, by a graph, and by solving an equation. What other ways can you use to predict when the two hikers will meet?

INTRODUCE

Expect to spend about 5 minutes.

1. Tie the situation to students' lives. If most of your students have cell phones, ask, ***Do you have a cell phone?*** Otherwise, ***Do you know people with cell phones?*** In either case, ***Do you know how cell-phone plans work? What do they cost?***
2. Open **Cell Phone.gsp** and go to page "Problem." Ask for a volunteer to read the problem aloud.
3. Probe to see whether students understand the problem. ***What might the data look like? Do you think the payments will all be the same?***

DEVELOP

Expect this part of the activity to take about 20 minutes.

4. Go to page "Plan" and press *Next Month* a few times. ***What do you notice?*** [Table values change.] Double-click the table to add the data for that month. Press *Next Month* and double-click the table again. ***Can we make a prediction yet?*** Keep clicking the button and double-clicking the table until the table has at least eight rows.
5. ***What might we do to make more sense of these data points?*** Elicit the idea of graphing them. Select the table and choose **Graph | Plot Table Data.**
6. ***How does this phone plan work?*** Add lines to the graph if students suggest doing so. Bring out the idea that the plan costs $30 for up to 300 minutes per month, with a charge for each minute over 300. ***Can we tell what the additional charge per minute is?*** Students might note from the graph or the table that each additional minute costs $0.40, or 40 cents. Mention that this amount is the amount the line increases vertically for each horizontal change of 1 minute. It indicates the steepness, or slope, of the line. Also show how this amount can be found from the table by dividing a difference of two costs by the difference of the corresponding times.
7. ***What alternatives other than this cell-phone plan might be considered?*** Encourage creative thinking beyond other cell-phone plans. A phone card that charges a fixed rate, though less convenient, might be more affordable.
8. ***How can we compare the cost of this phone plan to that of using a phone card?*** Suggest that we look at the average cost per minute.

9. Press *Plot Function* to show a graph that passes through the data points. Construct a point on the function plot and measure its coordinates. Choose **Measure | Abscissa (x)** and **Measure | Ordinate (y)**.

10. ***How can we calculate the average cost?*** [Divide the cost, *y*-coordinate, by the time, *x*-coordinate.] Use Sketchpad's Calculator and the coordinates to enter $\frac{y}{x}$, the average cost. Students can see the average cost change as you drag the constructed point along the function plot.

11. ***How does that average cost per minute change? When is it highest?*** [Close to 0 minutes] ***When is it lowest?*** [300 minutes] ***Does this make sense?*** Help students relate the cost per minute to fractions; as the time increases over the first 300 minutes, the numerator of the fraction (cost) stays the same, but the denominator (time) increases. After 300 minutes, both the numerator and denominator increase.

12. Go to page "Other Plans" and press *New Plan.* Demonstrate the process of generating a table by double-clicking the table, each time *Next Month* is pressed, plotting the table data, and plotting the function. When you press *Reset* or *New Plan*, the function plot disappears, but you will need to show students how to delete the table data and the points. To get rid of the points, drag the **Arrow** tool to form a rectangle that selects them all at once and press Delete on your keyboard. To get rid of the table data, select the table and choose **Number | Remove Table Data,** select **Remove all entries,** and click **OK.**

13. Let students investigate other cell-phone plans on page "Explore More." They can adjust sliders for the number of minutes, the fixed cost, and the cost for each additional minute.

SUMMARIZE

Expect to spend about 5 minutes.

14. ***How does the cost of the original phone plan compare to the cost of a phone card with a fixed rate?*** A phone card costing 10 cents per minute has the same average cost per minute as this plan if exactly 300 minutes are used; otherwise, the card has a lower average cost per minute. Explore some other fixed rates as well.

15. ***What have we learned through this investigation?*** Help students articulate whatever they have learned. Include the objectives of the lesson.
    - Graphs can help us make meaning of data in disorganized tables.
    - The slope (steepness) of a line is the amount of vertical increase for every unit of horizontal increase.

- The slope can be found from a data table by dividing the difference of a pair of second coordinates by the difference of the corresponding first coordinates.
- To compare two situations, we often want to find averages.
- Different graphs can represent the same situation.

16. ***What other questions might we ask?*** Here are some ideas students may raise.

    Why does a fraction decrease in value if the denominator increases and the numerator remains fixed?

    How do prepay cell plans compare?

    Would a cell-phone plan in which you paid more for more minutes really be cheaper?

    Is it always true that increasing both the numerator and denominator of a fraction gives a fraction of higher value?

EXTEND

If you choose to do the extension, expect to spend 15 minutes.

1. Students may suggest graphing the cost per minute. You can plot a single point whose x-coordinate is time and whose y-coordinate is the cost per minute, but the point will appear to lie on the horizontal axis, because of the scale. Suggest calculating a new value of 100 times the cost per minute and plotting this value as the y-coordinate.
2. This point moves more dramatically, and tracing it gives a graph that clearly shows the minimum value. With the point selected, choose **Display | Trace Plotted Point.** Then drag the point on the function plot. ***So, we have two different graphs that represent the same situation.***
3. You might pursue the question of why the average cost per minute in the original plan is increasing with minutes more than 300. The fact that the numerator and denominator are increasing could mean that the fraction is either increasing or decreasing. Students might say that the numerator is increasing faster than the denominator because 40 is more than 1. Others might argue that 0.40 is less than 1. If possible, elicit the idea that what matters is the relative amount of change. For the numerator to change from 30 to 30.4, for example, is a 13% increase. The denominator's corresponding change from 300 to 301 is an increase of only 0.3%.

www.ingramcontent.com/pod-product-compliance
Lightning Source LLC
Chambersburg PA
CBHW080706220326
41598CB00033B/5324

* 9 7 8 1 6 0 4 4 0 2 4 5 2 *